# Feeling the Heat

## Dispatches from the Frontlines of Climate Change

From the Editors of
*E/The Environmental Magazine*

Edited by Jim Motavalli

With contributions from Sally Deneen,
Ross Gelbspan, David Helvarg, Mark Hertsgaard,
Orna Izakson, Kieran Mulvaney, Dick Russell,
and Colin Woodard

Photographs by Gary Braasch

Routledge
New York and London

Published in 2004 by
Routledge
29 West 35th Street
New York, NY 10001
www.routledge-ny.com

Published in Great Britain by
Routledge
11 New Fetter Lane
London EC4P 4EE
www.routledge.co.uk



Routledge is an imprint of the Taylor & Francis Group.
Printed in the United States of America on acid-free paper.

10 9 8 7 6 5 4 3 2



Library of Congress Cataloging-in-Publication Data

Feeling the heat : dispatches from the frontlines of climate change / edited by Jim Motavalli
p. cm.
Includes bibliographical references and index.
ISBN 0-415-94655-7 (hc : alk. paper) — ISBN 0-415-94656-5 (pb : alk. paper)
1. Global warming—Environmental aspects—Case studies. I. Motavalli, Jim, 1952-
QC981.8.G56F42 2004
55.5'253—dc22 2003018461

# Contents


Preface vii

Introduction 1

Part One: Human Impacts

1. China: The Cost of Coal 11
   Mark Hertsgaard

2. Europe: Planning Ahead 25
   Colin Woodard

3. Greater New York: Urban Anxiety 39
   Jim Motavalli with Sherry Barnes

4. Antigua and Barbuda: Islands under Siege 61
   Dick Russell

5. Asia: Clouds Got in the Way 79
   Jim Motavalli

*Photo Essay*
Witness to a Warming World
Gary Braasch

PART TWO: ECOSYTEMS IN TROUBLE

6. Alaska and the Western Arctic: The Ice Retreats 95
KIERAN MULVANEY

7. The California Coast: Marine Migrations and the Collapsing Food Chain 111
ORNA IZAKSON

8. Australia, Florida, and Fiji: Reefs at Risk 127
DAVID HELVARG

9. Pacific Northwest: The Incredible Shrinking Glaciers 141
SALLY DENEEN

10. Antarctica: The Ice Is Moving 157
DAVID HELVARG

Author Biographies 173
Endnotes 177
About *E Magazine* 183
Index 185

# PREFACE

The reality of catastrophic climate change does not seem to be getting through to people, and it is not hard to understand why. Global warming is dismissed as speculation by the president and Congress, and cited—if at all—by the news media in confusing tit-for-tat exchanges full of scientific jargon. The small band of skeptics is given equal weight with the overwhelming majority of climate scientists, and their bantering about computer models, aerosols, and ice cores just confuses the public.

The cold winter of 2002 to 2003 was fodder for the morning shock jocks. "Where's your global warming now?" they asked. But the scientists are telling us that climate change is not simply a global hot foot; it is subtler and far more dangerous than that. Instead, we have entered an era of profound climatic instability, with more severe storms and great variations in temperature and rainfall. The broiling summer Europe endured in 2003 was a very dramatic example of the trend.

The essays in this book are reports from the climate front. As Ross Gelbspan notes in the introduction, the science of global warming is no longer being seriously debated. It is real, and it is here. From China to New York, minor changes in what were fairly established weather patterns have already produced profound and permanent effects on local ecosystems. Fish species are disappearing, with ripples throughout the food chain. Birds and butterflies are moving, turning up in places they have never been seen before. Some plants are dying, others thriving as manmade climatic changes accelerate.

It is no great mystery what causes global warming: Carbon dioxide emissions from coal-fired power plants, cars, and trucks; cooking fires; and deforestation result in most of it. Americans have a primary responsibility. The U.S. transportation system emits more carbon dioxide than any other nation's *entire economy* (with the sole exception of China), reports the Pew Center on Global Climate Change.

We could reduce those emissions with a collective will, embodied in legislation like the Kyoto Protocol, but very little progress has been made. In only a few places—most notably, Europe—are people not only paying attention but also acting on their awareness. (And even many of the European countries are missing their targets under the Kyoto Protocol.) Meanwhile, the ominous build-up of carbon dioxide in the atmosphere continues.

This book grew out of a lengthy collection of articles underwritten with the generous support of the Richard and Rhoda Goldman Foundation and appearing in the September/October 2000 edition of *E/The Environmental Magazine*. We wanted to move beyond the scientific debate. The idea was to document—through the kind of in-depth, heavily sourced reporting the magazine is known for—the evidence for a changing climate.

We certainly found it. Our reporters traveled to India, China, Australia, Fiji, Antarctica, and the Caribbean islands of Antigua and Barbuda. In the United States, we visited Alaska, coastal California, the Pacific Northwest, New Jersey, and New York City.

This book represents a considerable expansion of that original reporting, and adds a chapter on threats and challenges in Europe. Taken together, it offers overwhelming evidence that global warming is under way, producing exactly the extreme weather events predicted by the vast majority of the world's scientific community.

In his 1997 book *Hot Talk, Cold Science: Global Warming's Unfinished Business*, S. Fred Singer wrote, "At most, we believe there will be a modest warming in the [twenty-first century], generally beneficial for agriculture and human welfare. Available evidence suggests that none of the extreme fears about severe weather events, sea-level rise, and spread of diseases, is warranted."

But the "available evidence" suggests no such thing. The testimony in this book is reported from the field, through interviews not only with scientists but also with ordinary working people whose lives and livelihoods have already been profoundly affected by the very events that Singer speculates are unlikely to occur.

Is the choking cloud of particulate matter over most of Asia "beneficial for agriculture and human welfare"? Does the loss of beaches and rising waves help the vital tourism industry in island paradises like Antigua and Fiji? Is the disappearing ice in Alaska, which is already decimating keystone species like the polar bear, a "severe weather event"?

If the evidence in this book is not enough, consider these anecdotal news stories reported in 2002 and 2003:

- April 22, 2002: Glaciers are disappearing in South America. Within the next 15 years, all of the continent's small glaciers (80 percent of the total) will disappear, according to French glaciologist Bernard Francou. "The trend is so clear that you can't argue with the numbers," he says (*Grist Magazine*).
- November 9, 2002: Scientists are linking the loss of lobster populations in Long Island Sound to global warming. Dr. Alistair Dove of the State University of New York says that lobsters are dying from what the *New York Times* summarizes as the "stress of an environment that had become hostile to their ancient internal thermostats." According to Dr. Dove, "The correlation is very strong. Not proven, but strong. Climate is the killer here" (*New York Times*).
- December 9, 2002: The Arctic reports record ice loss, according to scientists from the American Geophysical Union. Surface melt in Greenland was the highest in recorded history. Arctic sea ice also reached a record low (BBC).
- December 11, 2002: The year 2002 will likely go down in history as the second warmest on record, exceeded only by 1998. "Studying [the] annual temperature data, one gets the unmistakable feeling that temperature is rising and that the rise is gaining momentum," says environmentalist Lester Brown (Earth Policy Institute).

- January 9, 2003: Dr. Andrew Derocher of the University of Alberta, Canada, says the polar bear could be driven to extinction by the loss of Arctic ice, which is melting at a rate of up to 9 percent per decade. Arctic summers could be ice-free by mid-century (BBC).
- February 14, 2003: In China, severe floods that used to occur once every 20 years now occur in 9 out of every 10 years. The number of people devastated by hurricanes or cyclones has increased eightfold to twenty-five million a year over the past 30 years. The oceans currently absorb fifty times more carbon dioxide than is contained in the atmosphere (*Guardian International*).
- February 26, 2003: Changes in forest productivity, the migration of tree species and potential increases in wildfires and disease could cause substantial changes to U.S. forests. The timber industry in the Southern United States is particularly vulnerable (Pew Center on Global Climate Change).
- March 7, 2003: Climate change effects could include a "big chill" for the Northeastern United States and Western Europe, with temperatures plunging as much as 9 degress F. The colder temperatures would be caused by the failure of the Gulf Stream to carry warm water from the tropics. The Gulf Stream makes an epic journey, traveling west over the top of Australia, around the Cape of Good Hope at South Africa's tip and up into the Atlantic Ocean. The moving water is cooled by the northern chill and becomes increasingly salty, sinking to lower depths for the return journey to the Pacific. This process has changed little since the last Ice Age, but global warming is throwing in a monkey wrench by melting ice in the Arctic Ocean. A UN assessment says Arctic sea ice in summertime could diminish 60 percent by 2050. This fresh water could dilute the salinity of the Gulf Stream, which would mean that it would not longer sink to the bottom of the ocean near Iceland and begin its return trip to the Pacific. According to Robert B. Gagosian, director of the Woods Hole Oceanographic Institution, "We're seeing huge freshening in the North Atlantic. The sinking of the cold, salty water has slowed 20 percent in the last 30 years" (*Wall Street Journal*).

- April 9, 2003: The Great Lakes region, which contains the world's largest source of freshwater, could face baking heat, droughts, floods, and other catastrophes as global warming raises its temperature over the next century, according to a 2-year scientific analysis coordinated by the Union of Concerned Scientists (Environmental News Network).
- May 28, 2003: Papua New Guinea is trying to convince two small communities of Polynesians, about two thousand people, that they should leave their homes on sinking tropical atolls northeast of Bougainville Island. Crops are reportedly being affected by seawater inundation. The Takuu people, one of the groups affected, have a 3,000-year history and more than one thousand unique songs. Eric Ani of Papua New Guinea's Disaster and Emergency Management Office says, "It probably is because of the effects of the greenhouse. There is talk of islands sinking everywhere in the world" (Agence France-Presse).
- July 3, 2003: The World Meteorological Organization (WMO), which usually produces technical reports and statistics, changed course to announce that the world's weather is going haywire. The WMO, which works with the weather services of 185 countries, documented record high and low temperatures, record rainfall and storms—and linked it to global warming. There were record temperatures in England and southern France, an unprecedented number of tornadoes in the United States, and severe monsoon heat waves in India. According to the WMO, 2003 could be the hottest year ever recorded (*The Independent*).
- September 9, 2003: The record heat wave that baked Europe in the summer of 2003 triggered forest fires, affected agricultural production, and proved deadly to thousands in France. "[Europe hasn't] seen such an extended period of dry weather and sunny days since records began [in about 1870]," said Michael Knobelsdorf, a German weather service meteorologist. In France, the heat wave claimed as many as 15,000 lives, many of elderly people. (Canadian Broadcasting Corporation/Planet Ark).
- September 23, 2003: The Canadian Ward Hunt Ice Shelf (which is up to 100 feet thick and had been in place for at least 3,000

> years) began to crack in 2000. In 2003, it broke in two, draining a trapped freshwater lake into the Arctic Ocean. Scientists attributed the disintegration to 100 years of relentless warming, a pattern that had accelerated in recent years. According to Dr. Warwick F. Vincent of Laval University, "The most recent changes are substantial and correlate with this recent increase in warming that we've seen from the 1960s to the present. It's an example where a critical threshold has been passed." Average temperatures in the Canadian Arctic have increased about four-tenths of one degree Fahrenheit every 10 years since 1967 (*New York Times/Washington Post*).

In sum, the evidence is clear that global warming is no longer speculative. Whether it is politically convenient or not, it has arrived. Controlling it is emerging as the major challenge of our time.

*Jim Motavalli*
*Norwalk, Connecticut*

# Introduction

In 1995, more than two thousand scientists from one hundred countries reported to the United Nations that our burning of oil, coal, and natural gas is changing the earth's climate. Nearly a decade later, many of the same researchers are very troubled by two things: The climate is changing much more quickly than they projected even a few years ago, and the systems of the planet are far more sensitive to even a very small degree of warming than they had realized. The average global temperature, the report said, will rise by 3 to 10° F by the end of the twenty-first century.

The accelerating rate of climate change is spelled out in two recent studies—one on the environmental side, one on the energy side.

In 2001, researchers at the Hadley Center, Britain's principal climate research institute, found that the climate will change 50 percent more quickly than was previously assumed. That is because earlier computer models calculated the impacts of a warming atmosphere on a relatively static biosphere. But when they factored in the warming that has already taken place, they found that the rate of change is compounding. Their projections show that many of the world's forests will begin to turn from sinks to sources—dying off and emitting carbon—by around 2040.

The other study, from the energy side, is equally troubling. Three years ago, a team of researchers reported in the journal *Nature* that unless the world obtains half its energy from noncarbon sources by 2018, we will see an inevitable doubling—and possible tripling—of atmospheric carbon levels later in this century. They followed that with another study, published in November 2002 in the journal *Science*, calling for a Manhattan-type crash project to develop renewable energy. Using conservative estimates of future energy use, they found that within 50 years, humanity will need to be generating at least three times more energy from noncarbon sources than the world currently produces from fossil fuels to avoid what would likely be a catastrophic build-up of atmospheric carbon dioxide ($CO_2$) later in this century.

Climate change is no longer a science issue. Stripped bare, it is a titanic clash of interests that pits the ability of this planet to support civilization versus the survival as we know it of the oil and coal industry, which is one of the largest commercial enterprises in history.

Unintentionally, we have already set in motion massive systems of the planet with huge amounts of inertia, whose stability has kept this earth relatively hospitable for the last 10,000 years. We have reversed the carbon cycle by about 400,000 years. We have heated the deep oceans. We have unleashed a wave of violent and chaotic weather. We have altered the timing of the seasons. We are living on an increasingly narrow margin of stability.

While climate science can be dizzyingly complex, the underlying facts are simple. Carbon dioxide in the atmosphere traps heat. For the last 10,000 years, we enjoyed a constant level of $CO_2$—about 280 parts per million (ppm)—until about 100 years ago, when we began to burn more coal and oil. That 280 has already risen to 370 ppm—a concentration this planet has not experienced for 420,000 years. It is projected to double to 560 ppm later in this century, correlating with an increase in the average global temperature of 3 to 10° F. (For perspective, the last Ice Age was only 5 to 9 degrees colder than the current climate.)

Evidence for the build-up of heat-trapping carbon dioxide abounds: The eleven hottest years on record have occurred since 1983; the five hottest consecutive years were 1991 to 1995; 1998 replaced 1997 as the hottest year on record; 2001 replaced 1997 as the second-hottest year and 2001, in turn, was replaced by 2002 in second place; the decade of the 1990s was the hottest at least in this past millennium; and the planet is heating more rapidly than at any time in the last 10,000 years.

The evidence, moreover, rests on a far broader base than computer models and temperature data. Add the unceasing bombardment of extreme weather events wreaking havoc all over the world.

Take 2001 as an example. At the beginning of that year, Britain emerged from its wettest winter in more than 270 years of record keeping. In early February, twenty-two successive blizzards in northern China stranded more than 100,000 herders, many of whom starved. In South Florida, the worst drought in 100 years decimated citrus crops, prompted extensive water restrictions, and triggered the spread of more than twelve hundred wildfires. In early May, some forty people died in the hottest spring on record in Pakistan. In June, Houston suffered the single most expensive storm in U.S. history when tropical storm Allison dropped 35 inches of rain in 1 week, causing $6 billion in damage.

In late July, a protracted drought in Central America had left more than 1.5 million farmers with no crops to harvest—and a million people verging on malnutrition. In Iran, a devastating drought resulted in more than $2.5 billion in agricultural losses. The drought was temporarily interrupted in August by Iran's worst flash flooding in 200 years that killed nearly five hundred people. In October, meteorologists documented a record ninety-two tornadoes in what is normally a quiet period for these events. In November, the worst flooding in memory killed more than one thousand people in Algeria. In Boston, after an October and November of record-setting warmth, it was 71° F on December 1—which prompted one observer to call this unseasonable weather "gift wrapping on a time bomb."

Why, then, is there any doubt in the public mind about the reality of climate change? And why is this *E Magazine* book necessary? Why send reporters to 10 global "hot spots"—from New York City to Fiji—for firsthand progress reports on the warming world? The answer lies in the millions of dollars spent by a shrinking number of industry players to maintain the illusion of "scientific uncertainty." Also to blame is the U.S. press, which, in general, has been too lazy to examine the scientific data and too intimidated by the fossil fuel lobby to tell the truth.

Even as villagers in Mozambique buried casualties of the horrendous rains that swamped the country in 1999, ExxonMobil declared in an ad on the op-ed page of the *New York Times*: "Some claim that humans are causing global warming, and they point to storms or floods to say that dangerous impacts are already under way. Yet scientists remain unable to confirm either contention." That flies in the face of the available evidence. But three years later, at the beginning of 2003, ExxonMobil remained a major funder of a tiny handful of industry-sponsored "greenhouse skeptics."

The Greening Earth Society, a creation of the Western Fuels Coal Association, takes a slightly different tack. Citing the opinion of a few greenhouse skeptics—most of whom are on its payroll—Western Fuels trumpets the idea that more warming and more $CO_2$ is good for us because it will promote plant growth and create a greener, healthier natural world.

These arguments neglect to mention that peer-reviewed science indicates the opposite. While enhanced $CO_2$ creates an initial growth spurt in many trees and plants, their growth subsequently flattens and their food and nutrition value plummets. As enhanced $CO_2$ stresses plant metabolisms, they become more prone to disease, insect attacks, and fires.

The impacts are even more negative in the world's tropical regions, where most of the poor and hungry live. As temperatures and carbon levels rise, plant biologists forecast a big drop-off in the rice yields in Southeast Asia, for example. A half-degree increase in the average temperature could cut India's wheat yield by 20 percent—this in a country where one-third of the population, more than three hundred million people, live in poverty.

Our knowledge about climate change comes from the work of more than two thousand scientists from one hundred countries participating in what is the largest and most rigorously peer-reviewed scientific collaboration in history. You would scarcely know this from the way the issue has been reported in the United States. The American news media have generally reported the issue as though the science was still in question, giving the same weight to the greenhouse skeptics as they do to mainstream scientists—all in the name of "journalistic balance." Real balance, reflecting the weight of opinion within the scientific community, would accord mainstream scientists about 95 percent of an article and leave a couple of paragraphs to the skeptics.

If the press really wanted to bring home the reality of climate change to viewers and readers, it would begin by making the connection between the warming atmosphere and the marked increase in extreme weather events—the floods, droughts, and severe storms that comprise an increasingly larger proportion of news budgets. Asked about the failure of the news outlets to make this connection, a ranking editor at one network replied, "We did include a line like that once. But we were inundated by calls from the oil lobby warning our top executives that it is scientifically inaccurate to link any one particular storm with global warming." The editor concluded, "Basically, our executives were intimidated by the fossil fuel lobby."

And resistance to the solution is staggering. We need to begin to move toward a global energy transition within this decade and we need to pursue it with the urgency of the Manhattan Project, in which the United States developed the atomic bomb in less than three years.

On one point the science is unambiguous: To allow the climate to restabilize requires worldwide reductions of carbon emissions of 70 percent.

The politics are almost as unambiguous. When 160 nations met in Kyoto, Japan, in 1997 to forge a climate treaty, not one government took issue with the science. Since then Holland has finished a plan to cut emissions by 80 percent in 40 years. In February 2003, British Prime Minister Tony Blair announced that the United Kingdom would cut its carbon emissions by 60 percent in 50 years. Germany has committed to reductions of 50 percent in 50 years. Even China, whose economy grew 36 percent during a 5-year period up to 2000, cut its emissions by 17 percent during the same period. In other words, the confusion around the climate issue stops at the boundaries of the United States.

The view of the world's business leaders is moving on the same trajectory. A vote by executives of the world's largest corporations, finance ministers, and heads of state who attended the World Economic Forum in Davos, Switzerland, in 2000 was remarkable. When conference organizers polled participants on which of five different trends were most troubling, the participants overrode the choices and declared climate change to be by far the most threatening issue facing humanity.

Climate change–induced damage to the Gulf Stream was a hot Davos topic in 2003.

Some of the world's largest oil and auto companies also acknowledge the perils of climate change and are positioning themselves for a new noncarbon economy. BP is now the world's leading provider of solar systems. Shell has created a new $1 billion core company to produce renewable energy technologies. Ford and DaimlerChrysler, together with Ballard Power Corporation, have entered a $1 billion joint venture to produce fuel-cell-powered cars. And both Honda and Toyota are marketing 50- to 70-mile-per-gallon climate-friendly hybrid cars in the United States.

The strongest corporate concerns about climatic instability come from the world's property insurers. During the 1980s, the insurance industry lost an average of $2 billion a year to damages from droughts, floods, storm surges, sea-level rise, and other extreme weather events. In the 1990s, it lost an average of $12 billion a year—$89 billion in 1998 alone. "Man-made climate change will bring us increasingly extreme natural events and consequently increasingly large catastrophe losses," an official of Munich Re, the world's largest reinsurance company, said recently.

But the impending damages go far beyond insurance losses. In early 2003, Swiss Reinsurance said it anticipates losses from climate impacts to jump to about $150 billion a year within this decade. Munich Re estimates that within several decades, losses from climate impacts will amount to $300 billion a year.

The largest property reinsurance company in Britain has projected that, unchecked, *the impacts of climate change could bankrupt the global economy within 65 years.* According to that scenario, the warming-driven damages from property loss, disease spread, crop destruction, and the disruption of infrastructures for energy, transportation, manufacturing, and public health—which are currently growing by about 10 percent a year—would exceed the world's annual gross domestic product in about 65 years.

While die-hard elements of the fossil fuel lobby continue to attack the findings of mainstream science, they are becoming increasingly isolated. For years, the Washington, D.C.–based Global Climate

Coalition (GCC) waged a campaign against mainstream science. But its corporate membership has hemorrhaged. At the end of 2000, the GCC collapsed after it was abandoned by Shell, BP, Ford, DaimlerChrysler, General Motors, the Southern Company, and Texaco.

The very few independent scientists who still question whether global warming is caused by human activity focus on discrepancies between satellite temperature readings in the upper levels of the atmosphere and on the ground. But those discrepancies were eliminated several years ago when researchers discovered that the satellite temperature readings were incorrect because scientists had failed to accommodate a natural decay in the orbits of satellites. When that decay was factored in, the satellite readings snapped into focus with ground measurements.

In fact, the argument of the naysayers has proved a moving target. Initially, the tiny band of greenhouse skeptics told us that climate change was not occurring. Then they told us it was so minimal as to be insignificant. Then they declared it is good for us. Recently, several of these skeptics have said, "Ooops, it's happening but there's nothing we can do about it."

Their position is mirrored by President George W. Bush who has declared that we will simply have to adapt to climate change. Bush has been especially antagonistic to the climate issue—perhaps because of his ties to the oil industry. Soon after his inauguration, he reneged on a campaign promise to cap emissions from power plants. He then released his energy plan—calling for thirteen hundred to nineteen hundred new power plants—which would be a prescription for climate chaos. Finally, Bush withdrew the United States from the Kyoto Protocol negotiations on the grounds that the treaty is unfair to the United States because it exempts developing countries from the first round of carbon fuel cuts. (Ironically, it was Bush's father, President George H. W. Bush, who approved that developing-country exemption back in 1992.)

But a number of commentators believe that, in addition to blocking U.S. action on the climate, the actions of George W. Bush may also be compromising America's traditional political role in the world. They argue that given the immensity of the climate issue and the imperative it is generating among other governments, Bush's withdrawal from the Kyoto process may be the first step in the transfer of global political leadership from the United States to the European Union.

Conditions are shifting rapidly, meteorologically and otherwise. Most of the public is now intuitively aware of climate change—and extremely worried about changes in the weather. Growing numbers of corporate leaders are realizing that the remedy—a world-wide transition to renewable and high-efficiency energy sources—would, in fact, create a huge surge of jobs and a dramatic expansion in the total wealth of the global economy. And national as well as grassroots political and religious activists are at last making the climate crisis the focus of campaigns. It is too slow and too small—but it is a beginning. The issue is not whether we will mobilize around the climate crisis. We have no choice. The issue is whether we will do it in time.

*Ross Gelbspan*
*Boston, Massachusetts*

# Part One

# Human Impacts

The scientific consensus . . . about human-induced climate change should sound alarm bells in every national capital and in every local community.

—Klaus Topfer, Executive Director, United Nations Environment Programme

# CHAPTER ONE

# China: The Cost of Coal

*Mark Hertsgaard*

My first morning in China, I was unexpectedly stricken with a fear known to all working journalists: Did I come all this way in search of a nonexistent story? Back in the United States, I had heard over and over again that China had the worst air pollution in the world, thanks to its overwhelming reliance on coal to fuel an economy that, throughout the 1990s, was growing by an average of 8 percent a year. All this coal burning was not only fouling China's skies, I'd been told, it had also made China the sec-

Fig 1: AP Photo/Greg Baker

ond-largest emitter of greenhouse gases in the world, trailing only the United States.

I bounced out of bed in Beijing that December morning and eagerly headed out for my first walk in the People's Republic. The temperature was a bracing 19° F, but the real shock was the brilliantly blue and sunny sky—not a hint of smog anywhere. How could this be?

It was just before 8 A.M. and the four-lane boulevard outside my hotel was crowded with a stream of humanity so dense and fast-moving that I could only stand back and watch. A few people traveled by car, rather more by bus, but the vast majority were on bicycles, usually the stolid black Chinese model called Flying Pigeons. There were also lots of three-wheeled cargo bikes whose wooden flatbeds carried everything from bulging sacks of fruit to freshly skinned sides of pork, to couches, televisions, and small mountains of crushed cardboard destined for recycling.

I was especially intrigued by the flatbeds I saw carrying the coal briquettes known as "honeycombs" (because of the holes drilled in the briquettes to encourage cleaner burning). Round, black, the size of small coffee cakes, the honeycombs were stacked by the hundreds into squat pyramids and sold off the carts for burning in the home stoves of the poor. Honeycombs were supposed to be the cause of much of China's pollution, but where was that pollution?

After a 45-minute walk around the neighborhood, I returned to my hotel chilled and bewildered: Beijing was grungy and strewn with trash, but it sure didn't live up to its advance billing as one of the most polluted cities in the world. That afternoon a government press aide proudly explained to me that the government had moved most of the city's heavy industry out of the downtown area. It sounded plausible, and I later learned that some factories had indeed been relocated. But I got the real story later that evening from a Beijing taxi driver.

It turned out that I had been witnessing a statistical fluke. The only time Beijing was graced by blue skies in winter was immediately after Siberian winds had roared through and flushed away all the smog; by chance, such winds had struck the night I arrived and continued blowing through the following day. As the taxi driver told me, the only reason Beijing's air did not look dirtier was that "it's very windy today. If there were no wind, you'd notice [the pollution] very strongly."

Sure enough, the winds calmed the next day, a Friday, and over the following 10 days I witnessed the sickening descent of Beijing into a city of murk and gloom.

At noon on Saturday, after just 12 hours of still air, I took a bus across town to a luncheon interview near Tiananmen Square. Straight above me the sky was still blue, but in the distance a pale gray layer of smog already frosted the skyline. When I came back outside 4 hours later, the layer had nearly doubled in thickness. The pollution accumulated with each passing day, and by Thursday I was used to waking up to a gray-white haze that rested on the skyline like a lid on a wok. The haze would grow palpably worse through the course of a day, as countless thousands of boilers were fired up and internal-combustion engines spewed exhaust.

On Thursday, I was riding south on the ring road that skirts Purple Bamboo Park on its way around the western edge of town. It was about 4:30 in the afternoon. My eyes should have been drawn to the Chinese national television (CCTV) tower, by far the tallest structure in the city, which lay directly ahead about 4 miles away. But by this hour the smog had become so thick that what had been a basically sunny day at noon now looked overcast and dark. Only because I knew the CCTV tower was up ahead could I faintly make out its needle-nosed outline against the sky.

Striding down the sidewalk to my right was a tall, young woman in a smart black overcoat. Behind her, two young girls in bright red and yellow athletic suits pedaled bicycles. A wizened old fellow in a blue Mao cap bent over an upturned bike, trying to repair it. All of us were inhaling lots of poison into our lungs.

On Friday morning, I took a taxi to the National People's Congress. Passing through the larger intersections of Beijing, I looked down the cross streets but could see no farther than half a block; beyond that, an impenetrable gray mass concealed everything. When I reached Tiananmen Square at 8:45, the sun hung barely visible above the southern gate to the Imperial Palace, like a weak light bulb in a barroom full of cigarette smoke. Gazing north, past Mao's mausoleum and the site of the 1989 massacre, I couldn't see the far end of the square, much less the Forbidden City beyond it. The pedestrians crossing the square were like spectral figures, half ghost, half flesh, as they disappeared into the gritty mist.

It had now been a week since Siberian winds had cleansed Beijing, and I craved a respite from the increasingly filthy air. On Sunday I took a public bus to the Great Wall, 46 miles north of the capital. The fresh air felt wonderful descending into my lungs, and northern winds even brought a cheering patch of blue sky by mid-afternoon. But not to Beijing. Back in the city, I stepped off the bus into a grimy dusk. Beneath my feet, whorls of coal dust spiraled across the sidewalk like black snow flurries.

I spent 6 weeks traveling throughout China in the winter of 1996 and 1997, and I found that the pollution in the other big cities, at least in the north of the country, was just as bad as in Beijing or worse. In X'ian, the ancient imperial capital known the world over for the collection of terra-cotta warriors buried outside of town, I conducted an experiment. Arriving by train at midday, I stepped out into the massive square outside the central station. It was a sunny day, but the pollution was so strong that the only sign of the orb itself was that one patch of sky was slightly brighter than the rest. I took out my watch and timed how long I could stare at that artificially veiled sun. Don't try this at home: The strength of the sun's rays can blind you. But in X'ian, it was no problem to stare directly at the sun for a full minute, so dense was the pollution.

## Bad Air, Bad Climate

A confession: Before coming to China, I had thought about its massive coal consumption mainly in terms of the implications for global climate change. But once I got there, I almost forgot to raise the point during some interviews I did with Chinese officials. When one is inhaling appallingly polluted air for weeks on end, one tends to focus the questions on *that*.

It is an important duality for outsiders to bear in mind, I think, when we consider the issue of poor countries and climate change. Outsiders tend to be most concerned with how coal consumption in China (or India, or any number of other countries) drives climate change—an urgent problem, to be sure, but one whose worst effects still lie in the future—while downplaying how coal burning is assaulting the health of people *today* in these countries. And the death toll is significant: Approximately two out of every seven deaths in China, according to World Bank figures, are attributable to the nation's unspeakable air and water pollution.

But climate change is likely to be a killer, too, and its onset appears inevitable. China's people and rulers are unlikely to be swayed from the path of rapid economic development and increased greenhouse gas emissions, even though China may already be suffering early consequences of global climate change, with yet more punishing effects likely in the years to come. In 1998, for example, floods roared through the Yangtze and Songhua River Valleys, leaving an estimated fifty-six million people homeless. Between 1999 and 2001, much of China experienced a record drought that reduced the grain harvest by 10 percent—an ominous development in a country that is already straining to feed itself and where some three hundred cities suffer from severe water shortages.

A study done by China's National Environmental Protection Agency, the World Bank, and the United Nations Development Programme concluded that a doubling of global carbon dioxide concentrations would have the following impacts on China: Storms and typhoons would become more extreme and frequent; much of China's coastline, including the economic powerhouses of Shanghai and Guangdong Province, would face severe flooding—with an area the size of Portugal inundated and an estimated sixty-seven million people displaced; food production would be affected most significantly of all, with increased drought and soil erosion lowering average yields of wheat, rice, and cotton; livestock and fish production would also decline.

These projections must be weighed against a second essential fact: In poor countries, common people and government officials alike see fossil fuel use as their ticket out of poverty. If I heard it once in China, I heard it fifty times: "We are used to it." That is what locals would answer when I asked whether the pollution bothered them. It took me a while to figure out exactly what they meant. They were not stupid; they did not enjoy breathing air that was so polluted it nearly glowed. Rather, what they were saying was that they were willing to put up with nasty pollution in exchange for warmer apartments and better pay, and finally having their own refrigerators.

"We have a saying in China," one Chinese journalist told me. "'Is your stomach too full?' In other words, are you so well off you can afford to complain about nothing? This phrase is used to describe Americans who talk about saving birds and monkeys while there are still many Chinese people who don't have enough food to eat."

My interpreter in China was a good example of this mind-set. Born in 1966, Zhenbing had grown up as the second of three sons in a small village near the border with Inner Mongolia. His family inhabited a mud straw hut and, like most rural Chinese of that time, was too poor to buy coal in winter; for heat, they burned straw and dried leaves. This, despite a climate as cold as Boston's or Berlin's, with winter temperatures that often dipped below zero. "Often the straw was not enough, so the inside wall of the hut became white with icy waterdrops, like frozen snow," Zhenbing recalled. "In my village, when a girl was preparing to marry, the first thing the parents checked was, is the back wall of the son-in-law white or not? If not white, they approved the marriage, because that meant his family was wealthy enough to keep the house warm."

These dreadful conditions began to change after Deng Xiaoping came to power in 1980 and implemented market reforms that revolutionized China's economy. Before long, money was trickling down to China's poor, and, not surprisingly, one of the first things they bought was more coal. White walls gradually became less common; at the age of fourteen, Zhenbing finally got his first pair of shoes.

Multiply the story of Zhenbing's family by the one billion plus people living in China and you understand both why China now has the world's most ravaged environment, and why no one who is serious about global warming can ignore the issue of global poverty. Over the past 20 years, average incomes have doubled in China, enabling hundreds of millions of Chinese to climb out of absolute poverty to "mere" conventional poverty. The environmental consequences have been as devastating as they were predictable. No one can begrudge the poor in

China (or anywhere else) a better life. Nevertheless, with 45 percent of humanity now subsisting on less than $2 a day, a difficult question arises: Can the world in the coming decades accommodate the ascent out of poverty of nearly three billion people without overwhelming the ecosystems that make life on this planet possible in the first place?

If only because nearly one of every four humans on Earth lives there, China will be decisive to our collective environmental future—so much so that it ranks as an environmental superpower. Like the United States, the other main environmental superpower, China wields what amounts to veto power over the rest of the world's environmental progress. China and the United States are each responsible for such a large share of global pollution that any attempts to, say, reduce greenhouse gas emissions simply cannot succeed without their cooperation.

The United States, with 286 million people, casts its environmental shadow largely through its extravagant consumption patterns; the average American consumes nearly fifty times more goods and services than the average Chinese. China's environmental heft still derives mainly from its massive population. But as incomes keep rising in China, the high-impact consumption patterns promoted by Western capitalism are becoming increasingly common too, and the consequences could be disastrous—not just for China but for all of us.

## A Mixed Picture

Yet China's environmental future is not necessarily as black as its skies are today. China has made impressive investments in energy efficiency over the past 20 years—much greater, in relative terms, than what the United States has done. If China continues this approach in the decades ahead, there is a chance that its production of greenhouse gases can be limited to an amount consistent with a living planet. It will not be easy—China's emissions are certain to increase even under the best of circumstances, and one must always approach the statements of Chinese officialdom on this matter (and most others) with caution. Nevertheless, the opportunity to "green" China is real, and Western governments and corporations could actually benefit politically and financially from this opportunity, if they have the wit to seize it.

Begin with the fact that coal is destined to remain at the heart of China's energy picture for many years to come. Why? Because coal is one of the few natural resources China has in abundance—China is the world's leading producer and consumer of coal—and because coal fuels much of the national infrastructure, particularly the electricity and manufacturing sectors. As of 2002, coal accounted for about two-thirds of China's total energy consumption and three-quarters of its greenhouse gas emissions. (The discrepancy stems from the fact that coal is, of course, the most potent carbon dioxide producer of the fossil fuels.)

From both an environmental and a public health standpoint, using less coal is clearly essential. Many Chinese might regard this as a recipe for economic stagnation, but the nation's recent history suggests otherwise. During the 1980s, as part of the economic reform program championed by Deng Xiaoping, China sharply reduced its rate of coal consumption by increasing its energy efficiency. Because the energy intensity of China's economy—that is, the amount of energy used per unit of Gross National Product (GNP) produced—fell by 30 percent during the 1980s, the economy was able to grow an average of 9 percent a year even as energy consumption grew by only 5 percent a year.

This remarkable accomplishment was the result of deliberate government policy, implemented in response to the global oil crisis of 1979. Energy planners realized that China could not hope to produce or buy enough energy to achieve the quadrupling of the Chinese economy by 2000 that Xiaoping wanted; the only option was to use energy more efficiently.

This was true because China's energy infrastructure was so antiquated: Many boilers and machines dated from the 1950s, and some from the 1930s, and their efficiency was very poor. To fix the problem, economic restructuring shifted the emphasis of China's economy from heavy to light industry. Subsidies and price controls that encouraged waste by making energy appear cheaper than it actually was were reduced. The introduction of more advanced boilers, fans, and pumps in the core electricity, steel, chemicals, concrete, and fertilizer industries further increased energy efficiency. More visible perhaps to the average Chinese was the phasing out of honeycomb home stoves—the least efficient, most polluting means of burning coal. This phase-out, which was still under way during my visit to China during the winter of 1996 and 1997, was pursued in concert with the demolition of vast tracts of *huotongs*, the traditional one-story courtyard dwellings of urban China, where residents invariably used home stoves for cooking and heating. The *huotongs* were replaced by high-rise apartment buildings that relied on central heating from industrial-sized boilers.

The improvements in energy efficiency continued and even expanded during the 1990s. Two of the leading individuals in the effort have been Zhou Dadi, a longtime senior official with the State Planning Comission's Energy Research Institute, and William Chandler of the Battelle Pacific Northwest National Laboratory of the United States Department of Energy. The increase in China's efficiency between 1980 and 2000, Zhou and Chandler have written, meant that the amount of coal needed to make a ton of steel fell by one-third by 2000; the amount of coal needed to make a ton of cement fell by 17 percent and a kilowatt hour of electricity by 8 percent.

Meanwhile, scientists from the Lawrence Berkeley National Laboratory at the University of California were helping to improve the efficiency of household appliances. This was crucial, because the rise of incomes in China was unleashing enormous demand for refrigerators, televisions, and other symbols of the global middle class. "In 1980," the Lab's China Energy Project website explains, "China produced fewer than 50,000 household refrigerators, 200,000 television sets, 250,000 clothes washers, and 13,000 air conditioners. By 2001, refrigerator and clothes-washer production exceeded 13 million, room air-conditioner production soared to 23 million, and more than 40 million color televisions were rolling out of factories, making China the largest appliance producer in the world. The increase in appliance ownership has led to an average 16 percent growth in household electricity consumption per year since 1980."

The new efficiency improvements promise to reduce the environmental burden of this surge in consumption. Although sales of appliances and consumer electronics items continued to grow at a double-digit pace through 2000, total household energy use was no higher than it had been in 1995. The Project estimates that, by 2010, China's new appliance efficiency standards will reduce projected residential electricity use by approximately 9 percent and greenhouse gas emissions by 56 million tons.

All this is good news—though not quite as good as certain analysts have suggested. In the summer of 2001, there was a flurry of media attention on the question of China's energy efficiency and its effects on global warming. In June of that year, the *New York Times* ran a story that credited China with having substantially reduced its annual greenhouse gas emissions during the late 1990s, at the same time that its economy was growing rapidly. The *Times* based its story, in part, on a report by the Natural Resources Defense Council (NRDC), which relied on data supplied by the Lawrence Berkeley Lab and the U.S. Department of Energy to assert that China had reduced its emissions by 17 percent from 1996 to 2000, even as its economy had grown by 36 percent. The NRDC report was issued at about the same time that the current Bush administration was condemning the Kyoto Protocol as bad for the American economy, and the implications the NRDC drew were clearly pointed at Washington: There was no necessary contradiction between fighting global warming and aiding economic growth.

But in August 2001, a *Washington Post* story suggested that China's performance was not as rosy as implied. The *Post* cited a Japanese researcher who had discovered that part of the reported decline in China's coal use (and therefore its greenhouse gas emissions) was bogus: Beijing had ordered many coal mines to close down, and local officials had reported they had been, but the mines had subsequently been reopened. It was an old story in China: The massaging of statistics to please bureaucratic superiors has long been routine, and local officials often pay mere lip service to Beijing's edicts, especially when their palms are greased to look the other way. "The mountains are high and the emperor is far away," goes the Chinese adage. For its part, the NRDC revisited the issue and, in a second report, found that greenhouse gas emissions fell between 6 and 14 percent from 1996 to 1999, while China's economy grew between 22 and 27 percent—still impressive, but considerably less so than originally claimed. In October 2003, the *New York Times* reported that new information from Chinese government agencies indicate that coal use "has actually been climbing faster in China than practically anywhere else in the world."

## New Energy

As China moves forward in the twenty-first century, efficiency remains the key to its energy future, the one bright spot in an otherwise gloomy picture. Because China began from such a low level of efficiency, there is still a long way to go—which, ironically, is good news for the environment. If the most efficient equipment and processes currently available on the world market were installed throughout China's energy system, China's energy consumption could be cut by 40 to 50 percent, according to studies conducted by Zhou Dadi and his colleagues at the Beijing Energy Efficiency Center.

To encourage such reforms, the World Bank has helped to create energy management corporations to spread the word about efficiency opportunities within China. The corporations approach factory managers with a simple message, said the Bank's Robert Taylor: "Managing your factory in an energy-efficient way will increase your bottom line. And if you don't believe it, here are some case studies that prove it." Unfortunately, the main thrust of the Bank's lending in China, as it is throughout the world, remains the encouragement of new fossil fuel development.

What about supply side alternatives to coal? Solar and wind power provide electricity in some sparsely populated regions in western China, but their impact on total energy consumption is miniscule. The same is true of nuclear energy, which supplies 1 percent of total electricity demand. Some Chinese planners have envisioned a bigger role in the future for nuclear power, but its high capital costs, to say nothing of its many safety problems, make that unlikely. (These drawbacks, however, did not prevent Chinese officials from announcing in 1997 that they would buy a $2 billion nuclear power plant from Russia.) Hydropower delivers about one-quarter of China's electricity, and total installed capacity is expected to quadruple by 2020. However, because electricity comprises only about one-fifth of China's total energy use, hydropower is destined to remain a small share of the overall energy mix.

Coal therefore is destined to remain a principal source of energy for China well into the twenty-first century, despite the ominous implications this carries for pollution in China and climate change globally. When I left China in 1997, it was widely expected to overtake the United States as the world's leading greenhouse gas emitter by 2020. Thanks to its acceleration of energy efficiency reforms, China now might not pass the United States until 2030. But either way, the additional carbon certain to be released is significant.

Nonetheless, Chinese officials make no apologies for having become a greenhouse giant. "Global warming is not on our agenda," a senior official of the Environmental Protection Bureau in Chongqing, the Yangtze River city that is perhaps the single most polluted place in China, said dismissively when I asked about his agency's strategies to reduce carbon dioxide emissions. As if to underscore his contempt for the issue, the official added an assertion he had to know was false—"All the pollution produced in Chongqing is landing here in Chongqing, so it's not a global problem"—before he declared, "We can't start worrying about carbon dioxide until we solve the sulfur dioxide problem." The official and his counterparts throughout China considered the sulfur dioxide problem more urgent because acid rain was landing on them and causing tangible damage *today*, while carbon dioxide emissions threatened merely potential, far-off, worldwide damage.

China has little patience with Western finger-pointing about climate change, regarding it as a means of constraining China's economic development. That is a paranoid view, but it contains a kernel of truth. For all its nuclear weapons, grand ambitions, and mobile phone-wielding businessmen (and women), China remains a poor country. On a per capita basis, it consumes only 10 percent as much energy as the United States, the chief culprit among the rich nations whose earlier industrialization had already condemned the world to climate change. So why, ask the Chinese, should we be held back? Should not the right to emit greenhouse gases be shared equitably among the world's peoples? To the Chinese, global warming is a rich man's issue, and if he wants China to do something about it, he has to pay for it. As one Western consultant with regular access to senior Chinese officials put it, "They know very well they can hold the world for ransom . . . and whenever they can extract concessions, they will."

"The Americans say China is the straw that breaks the camel's back on greenhouse gas emissions," comments Zhou Dadi, a self-described insider on China's climate change policies. "But we say, 'Why don't you take some of your heavy load off the camel first?' If the camel belongs to America, fine, we'll walk. But the camel does not belong to America. . . . China will insist on the per capita principle [of distributing emissions rights]. What else are we supposed to do? Go back to no heat in winter? Impossible."

Dadi, I should emphasize, is no party hack. He is fully aware of the prospects for global climate change, and doing what he can to prevent it through his tireless promotion of efficiency improvements. But he also recognizes the understandable appetite of the Chinese people for more comfortable lives after decades of want and repression. "China is not like Africa, you know, some remote place that's never been developed," Zhou told me when I interviewed him in a Beijing hotel one afternoon. Explaining China's goal of becoming what he called "a middle-class country" like France and Japan, Zhou added, "We used to be the most developed country in the world. Now, after many decades of turbulence, civil war, revolution, political instability and other difficulties, we finally have the chance to develop the country again. And we will not lose that chance."

Even leading Chinese environmental officials support the goal of continued, rapid economic growth. Few Chinese respect the Communist Party any longer. But the Party has kept the economy growing, and everyone is desperate to avoid a relapse into the kind of chaos, waste, and stagnation that befell China during the Cultural Revolution of the 1960s.

"This is the terrible dilemma of China's environmental crisis," argued one Chinese environmental advocate who must remain nameless. "Rapid economic growth is the most critical prerequisite for improving China's environmental situation. If economic growth stops, people will go back to the old, dirty, cheaper methods of production. Worse, there will be political instability, and that will overshadow everything; in that case, no one will have time to worry about the environment. Of course, this rapid economic growth will cause additional environmental damage; there are some things in the environment that are irreversible. That's why I think China will have to lose something—some species, some wetlands, something. We are working very hard to strengthen our environment. But, as much as I regret it, you cannot save all the things you would like. You cannot stop a billion people."

What you can do, however, is make sure that the growth is as green as possible—which is where Western institutions come in. The United States and other nations should follow what I call a Global Green Deal policy toward China: instead of investing in more coal, oil, and fossil fuel development—which remains the focus of most Western capital today—Western governments and companies should be selling and financing energy efficiency in China. Switching to more efficient light bulbs, insulating China's notoriously drafty apartment and office buildings, and installing more efficient electric motors in China's factories would not only decrease its dreadful air pollution and greenhouse gas emissions, but also provide much-needed jobs for workers and profits for companies in both China and the West. What's more, a Global Green Deal in China would not only help reverse the planet's deteriorating environmental situation, but also assist the climb out of poverty that is an inextricable part of humanity's struggle for survival in the twenty-first century.

## Chapter Two

# Europe: Planning Ahead

*Colin Woodard*

When the Dutch want to spend a summer's day by the sea, not a few drive up to the quiet village of Petten, perched on Holland's North Sea coast. Some stay in tidy cottages huddled within earshot of the roaring surf; others smell the sea air or watch seabirds from their hotel balconies or the windows of the handful of shops and restaurants near the hamlet's main intersection. But from town, nobody can get so much as a glimpse of the nearby sea.

As far as the eye can see, Petten's ocean view is blocked by a massive artificial wall, a 42-foot-tall earthen hill reinforced with stone, concrete, and steel. Only by climbing the stairs to the top of this seawall can one see the surf crashing near its fortified base. Looking down the seawall's ridge south of town, some visitors have a moment of dizziness, as their minds try to adjust to what their eyes are telling them. The houses, grass, and trees on the inside of the wall stand on ground that is lower than the head of the beach on the wall's exterior. On stormy days, the wall is all that stands between Petten and disaster.

With a quarter of its territory below sea level—and much of the rest threatened by coastal or river flooding—the Netherlands takes climate change very seriously. While many other countries have ignored scientific predictions that global warming will bring rising seas and changing rainfall patterns, the Dutch have been preparing themselves for the worst. Dutch engineers have raised the height of the Pettener seawall several times since 1976, doubling its height in an effort to stay ahead of storms, erosion, or rising seas.

Nationwide, the Netherlands plans to spend an extra $10 to $25 billion over the next century to upgrade dikes, pumping stations, and sea defenses. Dutch companies are even preparing to make big money helping other nations respond to what most here see as the inevitable consequences of pumping greenhouse gases into the atmosphere.

"It's better to be safe than sorry when you live below sea level," explains Peter C. G. Glas, director of inland water systems at Delft Hydraulics, the company that designs much of the country's extensive water management infrastructure. "We've had a tradition over the past century of being frightened of the water, and rightly so. We had the energy and the economic means to keep building up these dikes, so that's what we've done."

## Consensus for Action

When it comes to responding to the threat of global warming, Western Europe provides a striking contrast to the United States. While Americans remain generally ambivalent about climate change and take a wait-and-see attitude toward its potential impacts, Europeans regard the threat seriously and are taking concrete action both in their legislative bodies and on the ground. From the Netherlands to Venice, governments are spending lavishly to defend their countries against rising seas, while European energy and engineering concerns are leading the way in providing the world with technologies to deal with both greenhouse gases and the climate impacts they are believed to trigger.

"Here in Europe there is an absolutely overwhelming consensus amongst both scientists and businesses that climate change is real and we need to act upon this," says Alex Evans of the Institute for Public Policy Research in London. There are several reasons for this firmer commitment, according to Hermann Ott of Germany's Wuppertal Institute for Climate, Environment and Energy. In comparison with their U.S. counterparts, European corporations have a smaller influence over the European Union, Ott argues, and even large energy concerns such as Royal Dutch Shell and British Petroleum formally recognize the reality of global warming. "Europeans are more likely to accept that there are limits to growth and the carrying capacity of the planet," he says. "We've had to come to terms with a lot more competition with each other, and due to the limited space in Europe, environmental protection takes a much higher importance for the general population."

Nowhere is that more true than in the Netherlands, where 60 percent of the population lives below sea level, relying on a vast network of pumping stations to keep their homes from drowning with every downpour. Many countries have networks of canals and irrigation ditches to bring water to crop fields. In the Netherlands, the canals exist to get the water out. Stand beside the canal running past Glas's office at Delft Hydraulics and you will notice that the water level in the canal is one or two stories higher than the ground outside his building. Down the road, a large pumping station pumps water up to the canal from a low-lying drainage ditch collecting the runoff from a recent snowfall. Elsewhere in the country, enormous pumps push the water in the largest collection canals up to the sea, whose surf crashes 14 feet above the ground level of Amsterdam's Schiphol Airport. The Dutch have a saying that sums up an essential fact of life here: "Pump or drown."

Fighting rising seas is hardly a new undertaking for the Dutch. Their country has been sinking ever since Roman-era farmers began draining their marshy lands to plant crops. Those ancient Dutchmen soon discovered an unfortunate fact of geology: Drained land tends to settle and sink with the falling water table. To stop their lands from turning into swamps, the Dutch built dikes and canals around their fields and towns and used windmills to keep them pumped dry. This made their land sink even faster, so the Dutch built ever-larger dikes, canals, and seawalls to keep the water at bay.

In the twentieth century, the Dutch went on the offensive, diking and draining vast sections of the seafloor to create an entirely new province. A 20-mile-long dam was constructed across the mouth of the Zuider Sea, converting it into a freshwater lake, and thousands of square miles of shallow sea bottom were ringed with dikes and slowly pumped dry. It took years to convert the oozing mudflats into prime agricultural land, some of which is now tended by the grandchildren of the reclamation workers. Today a quarter million people live in the "New Lands" and another nine million Dutch citizens live in other parts of the country that lie below sea level.

Ironically, when it comes to global warming, the ultimate "low country" has little to fear from the sea itself. That is because in past decades and centuries, the Netherlands has built more than $2.5 trillion worth of water defenses, explains John G. de Ronde of the National Institute for Coastal and Marine Management in The Hague. "The fact that we have existing infrastructure makes it relatively simple to cope with sea-level rise," he says.

More than a decade ago, de Ronde's institute examined how various sea-level rise scenarios would affect the country, factored in expected increases in North Sea storm activity, and formulated a plan to address the dangers before they ever came to pass. Seawalls will have to be raised and strengthened. Major modifications will be made to the great dam at the mouth of the Zuider Sea, which controls the young lake's level by releasing water into the North Sea at low tide; if the world's sea levels rise significantly, the dam will no longer function properly.

It was de Ronde's team that determined the total costs of defending against the sea and rivers could run as high as $25 billion. "These are enormous figures if you had to spend them all at once, but we're able to spread it out over 50 or 100 years," de Ronde points out. (Over the course of this century, the Dutch might well spend that much maintaining their comprehensive system of bike paths.) The reinforcements and modifications around Amsterdam and other major cities are designed to withstand the combined effects of sea-level rise and the most powerful storm one would expect to encounter in a 10,000-year period.

What has Dutch engineers most worried most is not rising seas, but a change in the weather.

Climate models suggest that rainfall in this region of Europe may increase by 5 to 10 percent, while many glaciers in the Alps will melt away. That may not sound like much, but all that extra water represents a serious threat to the Netherlands, which sits smack in the middle of the Rhine River's flood-prone delta. Much of the meltwater of the Swiss Alps eventually winds up in Holland, as does most of the rain falling in Switzerland, western Germany, Luxembourg, and interior Belgium.

For centuries, the Dutch have built dikes to keep the river at bay. As their land continued to slowly sink, the Dutch have built their dikes higher and higher. But by doing so, the consequences of a breach of a dike have become ever more serious. In 1995, the Rhine almost breached its banks, forcing a quarter million people to evacuate their homes. The prospect of more frequent floods and ever-higher seas has given water managers pause. Maybe, many have concluded, building ever-higher dikes is no longer the best solution.

"There are flood plains that are inhabited that should not be," says Jeroen van der Sommen, managing director of the Delft-based Netherlands Water Partnership, which promotes Dutch expertise overseas. "We have to change our thinking and say, 'If you don't want to get your feet wet, you need to get out!'"

Rather than building levees higher to deal with global warming, the new strategy will give the rivers more room, allowing them to flow and flood more naturally rather than trying to force them into artificial channels.

Under the plan, 222,000 acres of land will be surrendered by 2050 to increase the size of the river floodplains, allowing them to revert to natural forest and marshland. Another 62,000 acres of pastures will be earmarked as huge temporary storage pools for floodwaters. Land-use practices on another 185,000 acres of farmland will be changed so they can tolerate soggy conditions in winter and spring. "It's a more complicated approach, but our future as a society will be much safer," says Peter Glas of Delft Hydraulics. "If the big flood comes isn't it better to have a meter (3.2 feet) of water in your house then to have six meters (19.6 feet) over your rooftop?"

But in the densely populated Netherlands, sacrificing land will not be easy. Some towns and villages will be told that they cannot build new infrastructure because their surroundings will be given back to the rivers in the coming decades. Dutch engineers are trying to find ways to minimize the dislocations, and even exploring ways to make use of areas that will in times of high water be allowed to flood. One company, Dura Vermeer, has even designed giant floating greenhouses, commercial parks, and towns that could be stationed in such places. In early 2003, the company was raising capital to build an enormous 123-acre floating greenhouse complex near Schiphol Airport, a prototype for future projects in other parts of the world threatened by similar global-warming-related flooding issues. "This could be the future for many countries," says van der Sommen.

Dura Vermeer is not the only European company that is adapting itself to a climate-altered future. Two huge European oil conglomerates—BP and Royal Dutch Shell—broke with their U.S. counterparts in 1997 when they pulled out of the Global Climate Change Coalition, a corporate alliance that has lobbied to block meaningful action on curbing greenhouse emissions. Shell, which is based in both London and The Hague, now says global warming is real and is investing $500 million in solar energy and biomass operations. London-based BP intends to cut its own greenhouse gas emissions by 10 percent from 1990 levels by the year 2010, twice the target agreed upon at Kyoto and later rejected by the United States. "We've moved—as the psychologists would say—beyond denial," BP chief executive John Browne said of the company's about-face at the time.

## Wind Power

The Dutch long ago abandoned their trademark windmills, wooden shoes, and the "milk maid" look in women's fashions. Clogs may never return, but in recent years wind power has been making a remarkable comeback. Huge, high-tech wind turbines spin in a long row mounted atop the great dike separating the "New Land" province of Flevoland from the waters of what was once the Zuider Sea. In 2002 alone, the Netherlands added 217 megawatts of new wind generators, increasing overall capacity by nearly a third, as part of its effort to meet Kyoto emissions targets.

Most of those turbines come from nearby Denmark, where the Vestas and NEC-Micon companies dominate the world's wind energy industry, which quadrupled its capacity between 1997 and 2002. These two Danish manufacturers supplied more than half the turbines now in use worldwide, which now represent one of the country's largest exports. In Denmark, wind turbines dot the countryside like gigantic pinwheels, while huge offshore wind stations harness the stiff winds from the Baltic and North Seas. The Middelgrunden Wind Farm just offshore from Copenhagen is the world's largest offshore wind plant, its twenty machines generating enough power to supply 32,000 households. In 2003, Denmark's wind turbines generated more than 15 percent of Denmark's electricity supply, and the government planned to increase that figure to 50 percent by 2030. Danish turbines have been exported to Germany and Spain, each of which has more wind-generating capacity than the United States. (The Germans have 12,000 megawatts of wind turbines, nearly four times the U.S. total.) "I don't think we can save the world with wind energy," says Christian Kjaer of the Danish Wind Manufacturer's Association in Copenhagen. "But we can show the world that being environmentally conscious doesn't have to come at the expense of economic growth."

Indeed, while much of the rest of the world has been talking about reducing greenhouse emissions, Denmark has actually been doing it. Authorities in central Copenhagen have deployed two thousand bicycles in public squares and train stations that can be borrowed for free, while a nationwide tax on automobile purchases more than triples the cost of buying a car. The residents of the windswept island of Aero receive much of their heat and power from the island's solar power station, which is the world's largest. Across the country farm, manure, and kitchen garbage are collected at special processing plants, which turn these wastes into fertilizers and clean-burning methane for some of the nation's power plants. These measures are helping Denmark meet its commitment to slash greenhouse gas emissions by 21 percent from 1990 levels by 2010, making it a world leader in addressing the causes of global warming.

## Fingers in the Dike

When it comes to dealing with the consequences of global warming, however, the world increasingly turns to the Dutch. In the future, many low-lying cities may wind up following the lead of Rotterdam, the great port city at the mouth of the Rhine in south Holland.

Because of its location, the Rotterdam region has always been vulnerable to flooding. But nobody realized just how vulnerable until February 1, 1953. Late that night a massive storm struck the South Holland coast, sweeping into the river mouths and smashing through dikes. A wall of water swept into South Holland and Zeeland, engulfing cities and towns. More than eighteen hundred people drowned. Fifty thousand homes and 200,000 head of cattle were swept away. The Dutch government vowed never to allow such a tragedy to happen again.

The solution was the Delta Project, the world's most daring flood protection system. In times of severe storm activity, a series of huge, movable gates can be raised to block all of the entrances to the river delta, including the mouth of the Rhine downstream from Rotterdam harbor. It took 45 years and more than $5 billion to complete, requiring the construction of several artificial islands, a 2-mile long movable storm barrier in the Zeeland marshes, and a pair of 60-foot-tall, 600-foot-long doors that can swivel shut at the mouth of the Rhine, protecting the city from another flood like the 1953 event. In normal times, the various gates and doors remain open, allowing rivers and ships to flow to the sea. Engineers compare the Delta Project to the building of the Great Wall of China or the Apollo space program.

With the threat of rising seas, many countries may consider a Delta Project of their own. In fact, Italy has been stumbling toward a very similar solution to save the world's most famous piece of sinking real estate.

## Venice under Siege

It is a clear, calm night—no storms, no wind, no unusual weather of any kind—but Venice's Piazza San Marco is flooding.

As the tide crests, seawater begins pouring from storm drains, forming a growing pond in the middle of the city's most famous square. The vestibule of the sumptuous San Marco Basilica begins filling with water, amusing tourists, some of whom are laughing as they wade through the ankle-deep pond.

But Venice's flooding is no laughing matter. The Adriatic Sea is rising, the city is sinking, and floods are becoming more and more commonplace. During the first half of the twentieth century, Venice was inundated by nineteen serious floods; in the second half, over 150. In 1996, Piazza San Marco, one of the lowest spots in the city, was flooded by high water on more than eighty occasions. The cumulative damage to the city's historic buildings, bridges, and art works is becoming increasingly apparent. Green algae now grows on the porous brickwork of many of the fourteenth- and fifteenth-century palaces along the Grand Canal because the flooding sea frequently tops the building's waterproof stone foundations.

Venetians have to replace their front doors regularly, as their bottom edges rot or corrode away. The ground floors of most buildings in the historic city have been abandoned for decades. Increasingly, the Venetians themselves are abandoning the old city, where the population has fallen from 175,000 after World War II to only 70,000 today. "People go away because it's more and more difficult to work and do normal activities," says Paolo Gardin of Insula, which works to reverse flood damage to the historic center. "We fear that the town is becoming only a museum for tourists."

Like the Netherlands, Venice has been sinking under its own weight for many centuries. Built on a series of swampy islands in the midst of a large lagoon, Venice's buildings had to be regularly replaced as they sank into the muck, and under its squares and walkways, archeologists have uncovered layer after layer of pavement laid over the centuries to keep residents' feet dry. But during the twentieth century, the rate of sinking accelerated, in part due to the pumping of vast amounts of groundwater by industry. Relative sea level increased by more than 9 inches due to a combination of sinking land and rising seas. Neglect made matters worse. During the late nineteenth and early twentieth centuries, Italians stopped maintaining Venice's canals and flood defenses, leaving the city more vulnerable to flood events.

As in the Netherlands, it took a terrible flood to awaken the nation's leaders to the threat. In November 1966, storm winds pushed waters up to the head of the Adriatic in the midst of a peak spring tide, swamping Venice under several feet of water for more than 24 hours. Phones and electricity went out as sensitive equipment slipped underwater. Large numbers of rats crawled up the walls of homes, while the canals were clogged with floating furniture and slicks of heating oil. There were no fatalities, but Venetians realized they could no longer live in ground-floor apartments. Something would clearly have to be done.

Unfortunately for Venice, Italian governments are less stable than their Dutch counterparts, and decades passed without any serious action taken on the soggy grounds of the Venice lagoon. "In Italy, governments change like the weather," says John Keahey, author of *Venice against the Sea.* "It takes many, many years to design and execute a project to protect the city, but Italian governments don't last more than a couple of years. It's the Italian way of political life."

As various flood control plans languished, scientists brought more bad news. The world's seas, they predicted, will rise considerably during the twenty-first century, raising the seas by as much as 3 feet relative to Venice's squares, streets, and low-clearance bridges. That could put much of the city underwater with almost every high tide. Slowly, haltingly, officials in far-off Rome finally moved toward approving an ambitious solution to Venice's watery predicament.

The plan, provisionally approved at the end of 2001, borrows heavily from the Netherlands' Delta Project. The centerpiece is a $3 billion project to build seventy-nine enormous hinged gates that can separate Venice and its lagoon from the Adriatic in times of flooding. The hollow steel gates, each of which measures about 65 feet to a side, will lie flat on the seafloor at the three entrances to the lagoon. When an *acqua alta* (literally, "high water") occurs, the gates will swing up to form a temporary wall, stopping the rising sea from entering the lagoon and the city's streets and squares. The gates would protect against flood surges of up to 6 feet in height, sufficient protection to keep pace with sea-level rise for at least 70 years, according to Giovanni Cecconi, an engineer with the New Venice Consortium, an alliance of major Italian engineering firms charged with designing, building, and operating the massive project. "It's not the final solution, it's just a way to protect the city during this century until another solution can come into place," he says.

Italian authorities have also started dredging canals, raising city pavements, and repairing damaged seawalls, slowly making up for a half-century of neglected maintenance. Italy expects to spend about $40 million a year for the next decade on such projects in an effort to buy the city some extra breathing room. "If you wish to defend the city from flooding, first of all you must maintain what is already there," says Paolo Gardin, who oversees much of the work. So far, construction crews have raised and restored many miles of canal walls and pavements. On the satellite island of Burano, home to some of Venice's famous lace makers, Insula built a series of small floodgates at all the island's canal entrances, allowing it to cut itself off from the lagoon in times of flooding. The only problem: the mini-gates provide protection only for floods up to 4 1/2 feet above normal water levels. The 1966 flood—and floods in a globally warmed future—stood more than 6 feet above normal water.

Thus exists the need for the big gates, say proponents of the massive project. "Virtually everyone agrees that the only ultimate solution is to separate the lagoon from the Adriatic," says Rafael L. Bras of the Massachusetts Institute of Technology, who led an engineering team that audited Venice's gates project. "Italy should do what the Dutch learned to do a long time ago: act now and think preventively," Bras says, rather than continuing to "take a wait and see attitude."

But the "big gates" project has drawn criticism from both environmentalists and some prominent scientists who warn it will turn into a financial and environmental boondoggle.

Environmentalists are generally skeptical of the New Venice Consortium, a private, for-profit enterprise that effectively serves as Venice's flood protection agency. "The Consortium is a little too focused on the gates project because that's what they live off of," says Paolo Lombardi of the Rome office of the World Wide Fund for Nature, which fears the project may consume resources better spent restoring the natural flood protection functions of the lagoon system.

But the gate's most prominent critic is Colgate University archeologist Albert Ammerman, a leading figure in Italian archeology. At various sites in Venice, Ammerman has literally uncovered evidence that the entire design of the gates project is based on false assumptions about the scale of the problem. "There are fundamental flaws in the scientific studies that the Consortium's gates are based on," he says. "They've really screwed this up."

The problem, Ammerman says, is that Venice is sinking much faster than anyone realizes. In archeological digs throughout the city, Ammerman's team has found layer upon layer of pavements and foundations laid over the centuries in a constant effort to keep the city above water. Through new carbon-dating techniques, the archeologists were able to use the layered pavements to calculate the rate of relative sea-level rise in the city over the past 1,600 years. The data, Ammerman says, show that the Consortium's plan seriously underestimates the likely rate of sea-level rise during the coming century. By adding Venice's long-term subsidence rate with the best available estimates of future sea-level rise, Ammerman reckons the sea will gain between 12 and 39 inches relative to Venice's streets.

If true, the sea still would not be nearly high enough to breach the massive gates when they are closed. The problem, Ammerman says, is that the gates will have to be closed far more often than their planners say they will. Since virtually all of historic Venice's sewers empty straight into the city's canals and lagoon with little or no treatment, frequent lagoon closures could have serious environmental consequences.

"In a bad year [in the mid-twenty-first century] you could have the gates closed for 100 or 120 days a year, with two-thirds of them in the three-month flood season," says Ammerman, seated at a terrace in Rome, where he spends much of the year. He predicts the resulting pollution crisis will shorten the project's lifespan to 30 or 40 years. "The building companies are happy to build something that's going to be obsolete in 50 years because then they can just build another," he says. The Consortium, he insists, should go back to the drawing board and revamp its designs.

Not surprisingly, that is not how the Consortium sees things. "The problem of sea-level rise is very real, and the designs take it into account," says spokesperson Monica Ambrosini. The gates will only have to be opened an average of five or six times a year a decade from now, she says, once all the improvements to canals and flood defenses are taken into account. The Consortium started construction in May 2003 with the support of Prime Minister Silvio Berlusconi. It expects to finish construction in 2011, but the process could be delayed by various legal challenges that remain.

Some supporters of the gates argue that the situation has become dire, and Venice can no longer wait. "If you want to preserve the city as a museum for tourists, you don't need to build the gates," says Andrea Rinaldo, a University of Padua engineer who has watched his native Venice lose nearly half its population since the 1966 flood. "But if you want a living city with real residents, decent jobs and everyday shops and stores then you must build these gates and protect the city."

Rinaldo remembers the 1966 flood, how his childhood home remained impregnated with oil months later. Since then he and many of his friends and neighbors have moved to the mainland. "People get tired of having to wait 40 minutes in a passageway for the tide to go down so they can walk home," he says. "You're afraid that you will die in the ambulance because [the ambulance boat] had to wait until the water got low enough so that you can pass under the bridge and on to the hospital. Have you even seen a grocery in town? It's all T-shirt shops now." The gates, he argues, will give city residents some certainty about the future and hopefully lead to its rebirth.

"They've protected Rotterdam from precisely the same threats, so why not protect Venice?" he asks, gesturing at a map of the lagoon on his office wall. “Because one thing is sure: If you let nature rule from this point then in 50 or 60 years there will be no Venice left at all.”

While Venetians may feel unprepared in comparison to the Dutch, they are far ahead of the United States (and much of the rest of the world) in their recognition that global warming is a threat that cannot be ignored. The Venetians and Dutch helped invent globalization, dominating international commerce back at a time when much of America had yet to be mapped. Perhaps it is not surprising, then, that they are ahead of the rest of us in confronting the threats posed by global warming and planning accordingly. “I do think that at some point the U.S. will take the lead for the world on these issues,” Ott says. “But for now, we Europeans are the countries where [parliamentary] majorities vote for the environment and are ready to plan ahead.”

CHAPTER THREE

# Greater New York: Urban Anxiety

## *Jim Motavalli with Sherry Barnes*

From a sea kayak floating off Pier 40 in lower Manhattan, you get a whole new perspective on New York City. The bustling metropolis falls away, and you are alone except for the sporadic barge traffic and the incongruity of students walking the high wire as part of a trapeze school in the Hudson River Park just beyond the seawall.

If the Hudson rises, it is most immediately noticeable to people like Randall Henriksen, who has led sea-kayaking expeditions here since 1994. From his perch in the front of the kayak, Hendriksen points to a green-and-white state Department of Environmental

Conservation sign on the seawall. "The algae there shows the mean high water line," he says. "It's been slowly but steadily moving up against that sign. The water has certainly been rising over the last few years, though you may not notice the change on a day-to-day basis," he says.

New York City, with more than seven million people, spills out over 378 square miles of land separated by the Hudson, East, and Harlem Rivers, Long Island Sound, and the Atlantic Ocean. The city, one of America's most diverse urban centers, is held together by a complex network of public works infrastructure, including roads, toll bridges, subway tunnels, water mains, gas lines, and millions of miles of telephone and television cables and electrical conduit.

It is a difficult city to run on a good day: In 1996, a "report card" prepared by the city's former U.S. Army Corps of Engineers chief gave New York's infrastructure failing grades, particularly for its aging water mains and solid waste treatment system, which dumps raw sewage into city harbors during storms.

So what happens when things get really bad? On December 11, 1992, a nor'easter storm hit the great city head-on. With wind gusts of up to 90 miles per hour and water surges 8 1/2 feet above mean sea level, New York's transportation infrastructure sputtered to a halt. Four million subway riders were stranded. The FDR Drive, the main highway along the east side of Manhattan, flooded up to 4 1/2 feet in some areas, and LaGuardia Airport, only 7 feet above sea level, grounded flights for the day. In the end, the federal disaster assistance totaled $233.6 million, according to Environmental Defense.

Was the storm a once-in-a-century fluke? Unlikely. Consider the summer of 1999, when high temperatures reigned over most of the eastern United States New York City experienced 27 days with temperatures of 90° F or more—double the number in an average year. Stores sold out of air conditioners, and 200,000 Manhattanites suffered a 19-hour blackout on July 7 because of excess power demand. Water consumption broke records, and thirty-three people died of heat-related causes in the city. The heat was accompanied by the worst American drought since the Dust Bowl of the late 1930s—rainfall in New York was 8 inches below normal for the summer.

But after the drought, a deluge occurred. Heavy rains soaked the city in late August that year, once again flooding the FDR Drive and the West Side Highway, and drowning some subway tracks in 5 feet of water. The big rainstorm was followed in September by Hurricane Floyd. The worst of the hurricane just bypassed the city, but total regional property damage was estimated at $1 billion. Since global warming brings with it the certainty of rising sea level and stormier weather, the city's aging infrastructure and delicate natural balance face unheard-of challenges.

Vivien Gornitz, associate research scientist at NASA's Goddard Institute for Space Studies, points toward a rectangular box jutting out of the Hudson in lower Manhattan, near a guarded U.S. Coast Guard booth. "That tide gauge uses an acoustic device to record the level of the sea's surface," she explains. "It takes a reading every six minutes." Gornitz and other researchers from Columbia University, New York University, and Montclair State University in New Jersey conducted an exhaustive study of the Metro East Coast (MEC) Region, which includes greater New York, Northern New Jersey, and Southern Connecticut, for the "U.S. National Assessment of the Potential Consequences of Climate Variability and Change for the Nation." The MEC findings were published by the Columbia Earth Institute in 2001.

One of the things that troubles Gornitz is all the recent construction at the water's edge. "Look, you can see it's on both sides of the river," she gestures, her arm taking in both sides of the Hudson just north of the former World Trade Center site. Gornitz fears that all the luxurious waterfront condominiums and commercial businesses are taking a risk that will increase dramatically as the new century progresses.

The most conservative climate change model used for the MEC study does not allow for rising greenhouse gas emissions; it merely projects the effects of the current rate of sea-level rise. By the end of the century, it says, we will be seeing 100-year floods every 50 years. "In the worst-case scenario, it could be as often as every four to five years," Gornitz adds. "It wouldn't mean the whole city under water, just the low-lying areas, including beach communities, coastal wetlands and some of the airports." And to further exacerbate the problem, the greater New York area is still experiencing land subsidence triggered by the glacial retreat that occurred more than 10,000 years ago.

New York City is not waiting for climate change: It is already experiencing much warmer years and reduced snowfall. Gornitz notes anecdotal effects, including the Central Park pond that people skated on in the 1970s that now often remains unfrozen all winter. “The cherry blossoms come into leaf a lot earlier now,” she adds, “and the leaves stay on the trees a lot longer in the fall.”

Janine Bloomfield is a senior scientist at Environmental Defense and author of the report “Hot Nights in the City: Global Warming, Sea-Level Rise and the New York Metropolitan Area.” Her report, based on MEC research, makes frightening reading. By 2100, she writes, New York City will have as many 90-degree days as Miami does today. “Sea-level rise will contribute to the temporary flooding or permanent inundation of many of New York City’s and the region’s coastal areas. . . . A large part of lower Manhattan would be at risk from frequent flooding by the end of the [twenty-first] century. . . . The East River would flood Bellevue Medical Center, the FDR Drive and East Harlem between 96th and 114th Street,” the report says. In a poignant note, the pre-9/11 report notes that the foundations of the World Trade Center would be vulnerable to nearly annual flooding at the end of the century. Droughts that now occur once in a hundred years could occur every 3 to 11 years by 2100.

“The tragedy of this is that we could do something about this now so the scenarios I wrote about won’t come to pass,” says Bloomfield, who now lives in Boston. “Unfortunately, we won’t react until the crises become obvious.”

The coming changes will do more than make people swelter or get their feet wet occasionally. “It really could become a serious economic burden for the city,” says Klaus Jacob, senior research scientist at Columbia University’s Lamont-Doherty Earth Observatory. “The current flood insurance program doesn’t account for 100 years from now, and that’s no way to plan for the future, especially a sustainable one.”

Coordinated planning for these eventualities has been minimal, and actual action even less. Some airport runways and seawalls have been raised. Rae Zimmerman, a New York University professor and director of the Institute for Civil Infrastructure Systems, complains that there is little cooperation between city agencies affected by climate change, and long-range planning is often the first thing cut from budgets that need to be slashed. Federal action has been nonexistent, with the Bush administration and Congress refusing to commit to anything more than redundant studies. But Jacob notes sardonically, "Whether Congress wants to address it or not, the sea level will rise."

## Losing the Bay

According to the *New York Times*, "Some have said that [Jamaica Bay's marsh] islands, rich with large and varied populations of birds and other wildlife, may largely disappear by 2020 if the causes are not found and remedies not applied." Cynthia Rosenzweig, a senior research scientist at NASA Goddard and also a professor at Columbia University, says the MEC project confirmed for the first time that Jamaica Bay's alarming wetlands loss is in part due to global warming. "Our wetlands researchers realized that something was happening out there that went beyond the usual stresses on this highly manipulated ecosystem," Rosenzweig said. "It's very complex, because there has been an interruption of sediment to the marshes, due to dredging for boat channels. Some people think the marshes are dying for reasons other than global warming, but we have documented with aerial photographs that climate change contributes to the loss by basically inundating the wetlands."

Jamaica Bay has been losing its marshlands at a rate of 3 percent a year since 1994, according to a Columbia University study, and 38 percent of marsh vegetation has disappeared since 1974. The construction of Kennedy Airport, built on marshland beginning in the 1940s, and other development was a major blow to the wetlands, though it remains one of the largest coastal ecosystems in New York State. The Bay was protected as the Jamaica Bay Wildlife Refuge in 1972 and became part of the National Park Service's Gateway National Recreation Area (which also includes New Jersey's Sandy Hook).

Federal protection has done nothing to prevent what appears to be an inexorable loss of land, which is dramatically illustrated in a series of aerial photographs of the Refuge's Yellow Bar Hassock taken in 1959, 1976, and 1998. "It is drowning," says Rosenzweig. Considerable biodiversity has been lost as well, including all the residents of what were once high marsh ecosystems there. In the dry words of MEC's climate change assessment report, "If Yellow Bar Hassock once had high marsh areas, as was suspected upon inspection of texture of some vegetation in the 1959 photographic print, then they were no longer in evidence during field visits." Jamaica Bay's ecosystem totaled 24,000 acres in 1900; by 1970 it was down to 13,000 acres. In the summer of 2003, workers began importing sediment and spraying it on Big Egg Marsh, trying to prevent its relentless shrinkage.

The borough of Brooklyn, now home to more than two million people, was once largely marshland, but the redesigning of this landscape for exclusive human use has resulted in the loss of valuable, natural protection in times of flood. "If you could imagine just putting a big sponge in front of lower Manhattan, that's what it would be like if there was a wetland there," explains Alex Kolker, a graduate student studying ecology and evolution at the State University of New York at Stony Brook.

One way to limit the loss of these flood barriers is to give coastal areas room to migrate inland. The U.S. Army Corps of Engineers (under the Clean Water Act) and the New York State Department of Environmental Conservation oversee current waterfront development. But according to Ellen Kracauer Hartig, a former research associate at Columbia's Center for Climate Systems Research, applicants can apply to bypass federal and state wetlands regulations, and permission is frequently granted. "At this time, the state gives out permits for development projects easily," she says.

That is an understatement. In 1998, the Corps rejected only 3.2 percent of major wetlands projects. Rejections are likely to become even more unlikely under Bush administration revisions that "streamline" the wetlands development process and relax Army Corps scrutiny in flood plains.

Global warming has also begun to affect the health of the city's residents. "In New York City, asthma rates in some neighborhoods are among the highest in the nation," explains Pat Kinney, an environmental health scientist at Columbia's Joseph L. Mailman School of Public Health. Kinney points out the well-established connection between air pollution, temperature, and rates of hospitalization and death. "What is new, is seeing how it all relates to climate change," he says, adding that raising the temperature in urban areas like New York, where there is limited vegetation to reflect heat and lots of concrete to absorb it, exacerbates health problems.

According to a 1996 American Meteorological Society report, an average of three hundred people a year die of heat stress in New York City. And there's a socioeconomic factor, too, explains Kinney: "Poor people, and especially elderly poor people, are most vulnerable to heat stress."

## The Virus Specter

Heat stress is probably the most obvious thing people think of when the subject of global warming comes up. Other effects are more subtle, but no less deadly. Higher rates of ground-level ozone are a major respiratory irritant, and vector-borne diseases thrive in warmer temperatures. And that is the problem that is keeping the city's public health officials up nights.

New York City had never had a case of West Nile encephalitis before 1999, but that hot summer—the hottest and driest in a century—sixty-two cases were reported in the region, and seven people died.

Tests at the Centers for Disease Control and Prevention in Atlanta and Fort Collins, Colorado, revealed that the illness was close to the St. Louis strain of encephalitis, which had never been previously reported in New York City. By September 6, there were five confirmed victims of the new virus and thirty-four suspected cases. By September 9, exotic birds began dying in the Bronx Zoo. A general health warning was issued, and city residents began to get used to helicopters overhead spraying clouds of malathion and pyrethriod pesticides. By September 21, scientists had isolated and identified the specific virus, not St. Louis encephalitis but West Nile.

West Nile is spread by a mosquito, *Culex pipens*, which breeds in stagnant pools of water. According to several prominent scientists, drought is the key factor in spreading West Nile virus. Outbreaks require an unfortunate series of events, they say. According to Dr. Dickson Despommier, a professor of public health at Columbia University, the mosquitoes' favorite prey is birds, but periods of high heat and drought send such common urban-dwelling species as crows, blue jays, and robins out of the city in search of freshwater. City bird populations are further reduced as unlucky individuals are bitten and killed by West Nile infection.

"By reproductive imperative the mosquitoes are forced to feed on humans, and that's what triggered the 1999 epidemic," Dr. Despommier says. "Higher temperatures also trigger increased mosquito biting frequency. The first big rains after the drought created new breeding sites." It took Hurricane Floyd, which passed through New York on September 16, to break the weather cycle that led to the outbreak.

Dr. Despommier says this same pattern is also discernible in recent West Nile outbreaks in Israel, South Africa, and Romania. In Bucharest, Dr. Despommier's investigation turned up abandoned buildings whose basements were full of water, a perfect *culex* breeding ground.

Another prominent proponent of the West Nile global warming connection is Dr. Paul Epstein of Harvard University. "Droughts are more common and prolonged as the planet warms," he says. "Warm winters intensify drought because there's a reduced spring runoff. The cycle seems to rev up in the spring, as catch basin water dries up and what's left becomes organically rich and a perfect mosquito breeding place. The drought also reduces populations of mosquito predators."

In 2002, the West Nile spread across the country, appearing in forty-four states and the District of Columbia. Five provinces of Canada were also affected. In a growing scientific consensus, public health officials believe the next drought will give this serious virus an even wider reach. Spraying certainly has not stopped these infectious bugs. Researchers at France's University of Montpellier said in mid-2003 that a mutation in the West Nile mosquitoes' genetic code resulted in their singular resistance to pesticides.

## NEW JERSEY'S BEACHES: ON SHIFTING SANDS

On stormy days, the wind at the tip of Fort Hancock, a former military base that is now part of the bustling Gateway National Recreation Area at the entrance to lower New York Bay, is enough to knock you down, and it churns the Atlantic into a froth favored by surfers but anathema to the embattled homeowners on this exposed coast.

Climate scientists predict that the sea level in New Jersey could rise an additional 2 feet in the next 100 years, with predictable havoc wrought on that priceless real estate. Beach erosion is likely to accelerate dramatically, too. But despite ominous reports of sea-level rise, and horrific damage caused by ever-increasing storms, proximity to New York City has meant rapidly escalating land values for this region, and a determination to build right to the water's edge. Even Fort Hancock, which can appear eerily deserted on a winter afternoon, is about to undergo a chic makeover.

Sandy Hook, where Fort Hancock is located, is like a finger pointed into the ocean toward Brooklyn, a beacon for the great New York/New Jersey estuary. The national park is a rare respite from a landscape dominated by beach communities and chock-a-block strip development. A former officer's quarters in the park, not far from nineteenth-century coastal defense emplacements, now serves as home to two organizations that are trying to protect this prosperous region from itself. The American Littoral Society and New York/New Jersey Baykeeper work together trying to preserve what is left of a natural environment laid low by dredging, filling, and construction.

Dery Bennett, the Littoral Society's friendly and grizzled director, takes visitors on a tour of nearby Sea Bright, where relatively modest vacation homes hide behind a protective seawall built in the 1930s. There is a 100-foot-wide beach behind the wall, built not over the millennia by the workings of the tides but beginning in 1996 by the U.S. Army Corps of Engineers as part of a $9 billion plan to "replenish" the beaches along the 127-mile New Jersey shore. The new sand is dredged from an offshore "borrow" site.

New Jersey is the poster boy for beach replenishment, since it is the only state in the union to pay its share not out of general funds but from a dedicated $25 million purse taken from realty transfer fees. Noreen Bodman, president of the business-oriented Jersey Shore Partnership, calls replenishment "a return on investment that benefits the state in terms of tax dollars, and ultimately benefits every resident in terms of quality of life and recreational values. It also protects businesses and utilities from the impact of some of these storms."

The luncheonette in downtown Sea Bright displays some starkly revealing aerial photos. One, taken in the early 1990s, shows a town with no beach to speak of, thanks largely to the effects of that seawall. The other, from 1999, shows a wide expanse of sand. The photos appear to offer stark proof that what human folly destroys, human ingenuity can repair. But many local environmentalists, including Bennett, Baykeeper Andy Willner, and Surfers' Environmental Alliance co-regional director Brian Unger, oppose the beach replenishment work. They say the massive effort to pump in sand benefits only a few wealthy homeowners, and also encourages even more dangerous shoreline development. And, they add, it is ultimately folly because global-warming-induced storms and rising tides will likely wash it all away in the next decade.

Orrin Pilkey's classic book *The Corps and the Shore*, written with Katharine Dixon, details how jetties, seawalls, groins, and other desperate maneuvers offer only temporary respite from the natural effects of erosion and shifting coastline—and eventually make things worse. The same thing is true of imported sand. New Jersey's replenished beaches, the authors wrote, could expect only a 1- to 3-year lifespan, at a cost of damage to coral, water clarity, and bottom-dwellers. In actual fact they have already outlived that prediction, though the sand is receding.

The East Coast was created in a collision between two tectonic plates, the American and Atlantic. Their coming together produced the Appalachian Mountains, and also the longest stretch of thin barrier islands in the world, extending from New England to Mexico. As Cornelia Dean notes in *Against the Tide: The Battle for America's Beaches*, "When these barrier islands are attacked by rising seas, their natural defense is to back out of the way." In other words, they are constantly shifting and reforming. Pilkey points out that barrier islands differ from any other topographic feature on earth because of "their ability to maintain themselves as a unit as they roll across a flooding coastal plain in response to a rise in sea level."

When this natural phenomenon meets global warming and the devastating effects of nonstop coastal development, rapid erosion is the result. "There's a natural process called littoral drift," explains Willner as he provides a pickup-based tour of Sandy Hook's windswept charms. "Sand from ancient granite mountains like the Appalachians was carried down by glacial action to create the beaches. Once here, it moves north in a predictable, inexorable fashion, reshaping the coast as it goes. What you see today is the result of millions of years of geological evolution, but people expect that process to stop when human infrastructure is introduced. They're putting homes and beach clubs on mobile land. And they're taking a crapshoot that those natural processes won't happen in their lifetimes. When it does, they're always surprised."

The speed with which the ocean reclaims its own is exacerbated by rising tides. According to Norbert Psuty, a coastal geomorphologist with the Department of Marine and Coastal Sciences at Rutgers University, deeper in-shore waters means more powerful waves, which move more quickly and retain more energy. In the last 100 years, the New Jersey coast has sunk 16 inches, through a combination of tectonic plate depression and sea-level rise. "Almost everything we have along the coast is at risk sooner or later," says Psuty. "We've been fortunate not to have taken any direct hits lately." Stephen Leatherman, who directs the Hurricane Center at Florida International University, puts it another way: "The erosion rates are going to accelerate in the future, which means the cost is going to go up exponentially to maintain these beaches. And no one seems to have figured it out yet. It's like a great big secret."

## Subsidized Privacy

There are no easy answers on the Jersey shore. According to the *Philadelphia Inquirer*, property that was worth $8.7 billion in 1962 is now worth $34.3 billion when adjusted for inflation. In 1945, George

Lippincott bought a house with 1.2 acres in coastal Avalon for $500, raising the money by selling a single rare stamp. In 2000, Lippincott's descendants put the property on the market for $3.5 million. The coast is now fully developed, with the result that a "100-year storm" would be far more devastating today than it would have been 50 years ago. Taxpayers will foot much of the bill for any rebuilding, since flood insurance is federally guaranteed.

The public trust doctrine, derived from English common law, says that states hold lands under tidal and navigable waterways in trust for their citizens. The concept has been incorporated into many state constitutions, and is generally interpreted as guaranteeing public access to shorelines up to the mean high tide mark. The town of Greenwich, Connecticut, fought a long and ultimately losing battle to maintain the exclusivity of its beaches that went as far as the State Supreme Court. It began when local attorney Brenden Leyden was turned away from jogging at a Greenwich beach, and it continued for 6 years. Fortunately for citizens not lucky enough to live in one of the United States's wealthiest towns, the public trust and First Amendment (claiming that the beach is a "traditional public forum") arguments eventually prevailed.

What does global warming have to do with beach access? Quite a lot, actually. The northern New Jersey coast is now mostly in private hands, and the public has only limited access to surf and sand. The scene is set for self-interest. The property owners who benefit the most from beach replenishment use their political clout not to enrich the shoreline commons, but to protect their own land values, sometimes with the active assistance of community leaders.

The I'm-in-it-for-myself mentality dictates more privately built jetties and seawalls, which accelerate the erosion damage caused by rising sea levels. And it means security guards and high fences on what was once open shore. Meanwhile, the public, by the very fact of their exclusion, loses its interest and its stake in protecting a coastal resource it can only see through locked gates. Sixty-seven-year-old Sea Girt resident Bob Devlin told the *Philadelphia Inquirer*, "I gave up going to the beach there a long time ago."

Public access and beach replenishment collide head on in Long Beach Township communities such as Loveladies and North Beach. To be eligible for federal funding, the towns are nominally required to provide open beach access every quarter mile, but endless rows of closely built houses, without the "street ends" that allow parking and foot traffic, dictate that the actual distance between access points is more like a mile and a half. And even where access does exist, the scarcity of parking (in some cases by design) limits its value to out-of-towners.

One of the groups that have suffered both because of beach replenishment and lost public access is the surprisingly strong northern New Jersey surfing community. The attraction is clear: It is state-of-the-art surfing almost within sight of New York City. As Surfline.com points out, "Sandy Hook boasts one of the few point breaks in New Jersey." Brian Unger, a graying but fit surfer turned environmentalist and access activist, takes visiting journalists on a tour through some of the exclusive beach towns near Sandy Hook that benefit from both beach replenishment and storm insurance, but make it as difficult as possible for the taxpaying nonresident to enjoy the imported sand.

The tour began on a blustery day in Elberon, an exclusive section of Long Branch just north of Bruce Springsteen's Asbury Park. Surfers fear that a pending beach replenishment in Elberon will smooth out the beach, remove natural rock formations, and affect surf-friendly wave formation by bringing deeper water closer to shore. They are arguing for a more nuanced approach that might use offshore reefs, different sand designs, and a much more gradually sloping underwater contour than the Army Corps of Engineers and the New Jersey Department of Environmental Protection had planned.

Elberon was once an ocean resort town for U.S. presidents and known as the Hamptons of the nineteenth century. James Garfield died there in 1881, and the spot is marked with a plaque. The Church of the Presidents, summer worship center for Presidents Ulysses S. Grant, Rutherford B. Hayes, James Garfield, Chester A. Arthur, Benjamin Harrison, and Woodrow Wilson, is now in disrepair, but it remains in a very upscale neighborhood.

It is unlikely the presidents were drawn by Elberon's great surfing, but they would have had no problem getting to the water if they wanted to try out a board. Today, it is far more difficult. In nearby Deal, summer home to many wealthy Sephardic Jews, huge estates have names like "Chez Fleur" and "Belle Mer." The Deal Casino Beach Club offers free parking but is restricted to residents. Meanwhile, the police are kept busy writing tickets on the nearby beach streets, where 2-hour limits are strictly enforced.

Many streets that once ended in public beach access are now off-limits, Unger says, because the municipality sold off the street ends to homeowners (a practice that was stopped only after state intervention). Despite the exclusivity, Deal is also slated for federally subsidized beach replenishment.

It is probably safe to say that wealthy property owners want to limit the invasion of young surfer kids and grizzled fishermen with their bait buckets and six packs of Budweiser. Many immigrant families from Newark or Paterson cannot afford the $5 and $8 daily fees at the lifeguard beaches, so some wait until late afternoon when the money collectors, mostly college students employed for the summer, leave for the day.

The few remaining free public access points between the million-dollar homes are hard to find and fairly forbidding. Unger led the way down a dangerous pile of construction debris that is the only public entry point to one lovely stretch at Darlington Beach. "Attention: Unprotected Beach. No Swimming," reads a sign. It was plain we were not welcome, but it was also plain that this nearly empty stretch of contested sand was worth the effort we made to reach it.

The Jersey shore town of Point Pleasant Beach developed a particularly bad reputation for harassing beach users in the 1990s: Surfers were told to get out of the water by private security guards, and people walking along the high tide line were ordered to leave the "private beach." Curbs were painted yellow to deter would-be parkers. The residents even posted signs that proclaimed: "Private Property. No Trespassing" (followed by, in tiny letters, "When Beach Is Closed").

But in 2002, spurred by the local activism of groups such as Citizens Right to Access Beaches (CRAB), the State Attorney General's office stepped in and forced a settlement that opens the entire beach "from the water to the edge of the dune" to the public. "The case law is very advanced," says Deborah A. Mans, an attorney for the New

York/New Jersey Baykeeper. "There has to be access to the mean high tide line, and as intervenors in these cases we're asking for 30 feet above that."

"The homeowners are just trying to make it as hard as possible," says Unger, who has run for the State Senate on the Green Party ticket. "But at some point you have to take a philosophical stand and say, 'No, I won't buy a beach pass because the beaches belong to the people.' But from Deal to Sandy Hook you have to really work hard to get on the beach without paying."

## Shifting Sand

With no access to the municipal beach and no parking even if there was, Unger sits in his van and points to the surf as it crashes on the beach. "We've already lost 30 feet of beach since they replenished this stretch a few years ago. It's something that we'll have to keep doing, and it's galling that for the most part it benefits only a few wealthy property owners."

But dire predictions that all the replenished sand would be washed out to sea within 2 or 3 years have not yet been borne out. "It's staying longer than I thought it would stay," admits Dery Bennett. "But everybody knew from the beginning that it's only a temporary fix. There will probably be five to seven replenishments within the next 50 years."

In his office at Sandy Hook, Bennett points to a chart of the Jersey shore. "Do you see how skinny those barrier islands are? Because of global warming there's faster erosion and these barrier islands are drowning. The Army Corps of Engineers ignores the accelerating of sea-level rise because some of them believe in a flat earth."

The Army Corps does not, in fact, officially ignore sea-level rise. Anthony Ciorra, project manager for the Army Corps Beach Erosion Control Project that stretches from Sandy Hook south to the Barnegat Inlet, said in 2003 that all but a few miles of the 21-mile replenishment project is completed. "We definitely factor in sea-level rise as part of the project design," Ciorra says. We calculate what it has been over the last 50 years for our benchmark."

But, of course, most scientists believe that sea-level rise will accelerate dramatically, making a hash of calculations based on the historic precedent. The Metro East Coast report, for instance, says that beach erosion is likely to double by the 2020s, increase by a factor of 3 to 6 by the 2050s, and by a factor of up to 10 in the 2080s. Nonetheless, Ciorra is triumphant. "I think we've been vindicated," he says. The project has exceeded our expectations. Our non-federal sponsors, the state of New Jersey and the local towns, consider it a success and are pleased with our performance."

In places where beaches have disappeared, it is sometimes hard to isolate a single culprit. Rising tides and severe storms, known global warming effects, work together with the ravages of development that stops only at the water's edge. Scott L. Douglass, author of *Saving America's Beaches* and a professor at the University of South Alabama, worked his way through college lifeguarding on the Jersey shore. Like many beach experts, he worries about higher sea level, but he is also a major critic of the erosion-promoting effects of jetties, seawalls, and dredging. Human activity has removed "more than a billion cubic yards of sand from the beaches of America, enough to fill a football field over 100 miles high," he points out.

Still, Douglass is relatively bullish on beach replenishment. "Replenishment adds sand to the system," he says. "It's positive if done correctly. But whether it's a long-term answer is a good question. We know that sea level has been rising at a rate of six inches per 100 years, and our beaches have kept up with that. But will they be able to keep up with the serious erosion problems caused by an increased rate? There's a lot of uncertainty."

Because we know the tides will rise, Douglass says the smart thing to do is "eliminate avoidable sand loss." But that is a difficult concept to embrace for homeowners facing the loss of their valuable real estate to relentless tidal action. Jetties and seawalls *do* protect owners' property with a temporary fix, but they also wreak havoc and steal sand from their neighbors.

Development along the water's edge is never-ending, pushed by the rising tide of real estate prices. Tourism brings in $12 billion annually to the New Jersey shore. In Sandy Hook, Bennett's stand against beach replenishment and unfettered development has put him at odds with the business community, which sees both as vital to its busy tourist season. Real estate lobbyist Ken Smith calls Bennett "a lousy misanthrope" for opposing more coastal development, and Bennett returns the volley by labeling Smith "a shill for the real estate industry."

Much as the environmentalists might want it, there is no groundswell for a retreat from the shore, the option favored by critics like Pilkey. The radical concept evokes both amusement and anger from New Jersey property owners and businesspeople. What, just walk away from billions of dollars of investment, not to mention the emotional attachment?

But Pilkey, who directs the Program for the Study of Developed Shorelines at Duke University, says retreat from the shore at least got a hearing in North Carolina, where he lives. He cites current events on North Carolina's Outer Banks, where the Dare County Project has actually concluded that moving buildings back from the water's edge would be cheaper than a program of 10 to 20 years of constant beach nourishment.

In New Jersey, the beach loss has so far been gradual and in many cases reversible. But they have good reason to fear the sea's restlessness in North Carolina, where global warming is helping to rapidly reshape the shoreline. The more than 130-year-old Cape Hatteras lighthouse was moved a half-mile inland in 1999 because a National Research Council study had shown that the shoreline in front of it was due to retreat 400 feet by 2018.

"Buying all the beachfront buildings on the Outer Banks would be cheaper than nourishing the beach, but they decided to nourish anyway at a cost of millions per mile," says an exasperated Pilkey. "Twenty years ago retreat from the shoreline wasn't even mentioned in the cost-benefit analysis, but now it's not so outrageous and at least has to be considered as a real possibility. We need to understand that when we build a high rise on the beach, we're forcing generations of people to defend it against the rising ocean."

Like most observers who closely follow coastal New Jersey, Pilkey cannot say how much of the rapid erosion there is caused by global warming. "Those beaches are very far north, so the nor'easters produce a great deal of wave energy," he says. "It's the longest stabilized coast in the U.S. Some of the beaches have had seawalls at their back for a long time, more than 100 years, and the effect has been to deepen the shore face and make the problem worse. So it's hard to separate global warming from all the other things that humans have done to the beaches. But there's no question that climate change is a factor. Global warming has to increase coastal erosion."

Pilkey was one of a group of eighty-five coastal geologists who wrote to President Ronald Reagan in 1982 urging "a new approach to the management of the American shoreline." As cited in Cornelia Dean's *Against the Tide,* the scientists had concluded that most shoreline stabilization projects protect property, not beaches, and that over 10 to 100 years these efforts "usually result in severe degradation or total loss of a valuable natural resource, the open ocean beach."

In their book *The Beaches Are Moving,* Pilkey and Wallace Kaufman point out that the barrier islands protecting the Gulf and East coasts are constantly on the move. These "warehouses of sand" have retreated many miles since they were created. The islands were fed by sand pumped forth by ancient rivers that in many cases have stopped playing such a role.

A 1990 *Coastal Management* article by James G. Titus of the Environmental Protection Agency entitled "Greenhouse Effect, Sea-Level Rise and Barrier Islands" uses New Jersey's 18-mile, two- to four-block-wide Long Beach Island, 15 miles north of Atlantic City, as a case study. It is an island of single-family homes clustered in such evocatively named beach towns as Surf City and Barnegat Light, and it is in danger.

Barrier islands' response to sea-level rise, says Titus, can be either to roll up landward ("similar to rolling up a rug"), in which case it remains intact, or it can break up and "drown in place." Titus's study envisions an accelerating, six-inch sea-level rise affecting Long Beach Island between 1986 and 2013, and another 6-inch rise between 2013 and 2031. Without coastal protection, barrier islands such as Long Beach are likely to simply become uninhabitable. To prevent that, Titus imagines residents will eventually have to approve a $1 billion, $219-per-household "keep things as they are" scenario of raising the entire island in place. Such an approach is likely to win more support than an engineered retreat from the shore, which would involve abandoning homes and buildings too close to the shoreline. But even $1 billion is not too high a price when $20 billion—a century's worth of fair-market rents—are at stake.

## The Big Makeover

Sandy Hook's quiet Fort Hancock looks like the one place northern New Jersey's developers forgot, but appearances are deceiving—plans are well under way to turn time-forgotten Fort Hancock from a quiet corner into a bustling conference center.

The developers, Sandy Hook Partners, share rent-free office space at Fort Hancock with the Jersey Shore Partnership, which is perhaps the biggest civic booster for beach replenishment. James Wassel, president of the Partners and of the larger Wassel Realty, has the kind of self-confidence that comes from a lifetime of standing in front of skeptical town boards and showing them plans for big buildings. A veteran of the Rouse Company (creators of Faneuil Hall in Boston and the South Street Seaport in New York) and commercial realtor Cushman Wakefield, Wassel insists he is not supporting the kind of big-ticket mall development that his résumé might suggest.

Wassel makes historically informed presentations even when his audience is only one wet reporter with a notebook. "This property was an Indian reservation in the early 1800s," he said. "A lighthouse [now the oldest continuously operating lighthouse in the U.S.] was built in 1764. The military started using it as a proving ground for new weapons in 1870. They used to put dilapidated ships offshore and blast away at them to test the range and accuracy of their guns." In the 1890s, as those guns developed longer ranges, Fort Hancock became the first line of defense for New York City.

The fort sits on 140 acres, with 110 buildings still standing. Sandy Hook Partners plans to spend $80 to $90 million rehabilitating the fort properties, though its agreement with the National Park Service means it cannot build so much as one new taco stand. Still, some of the dilapidated buildings will become gleaming restaurants and quaint inns, complete with manicured lawns, and some people are objecting to it.

"This is the last undeveloped stretch of shoreline in New Jersey," says Brian Unger. "I don't think it needs conference centers, restaurants and all that stuff." Cindy Zipf of Clean Ocean Action worries about a public space becoming private, "even though the developers say they won't change a hair on the buildings' chinny chin chins. The pressure to make money will be huge, and we don't want a multi-million dollar mogul to repair buildings and turn the place into a mini–Woods Hole."

But while most local environmentalists would probably prefer for the fort to remain wild and free, the buildings are crumbling rapidly and need emergency intervention. With only $250,000 in annual federal funding, the Park Service estimates that within 5 years many of the historic buildings at Fort Hancock "would likely deteriorate to a condition beyond repair."

Given the development restrictions, what Wassel and his colleagues envision is not a nautically themed mall but an environmentally oriented learning and conference center that would attract corporate clients interested in, among other things, the effects of global warming on coastal America. Instead of Starbucks, there will be low-key bed and breakfasts. It may open for business in 2008.

Wassel does not seem too concerned that flooding is a regular headache at Fort Hancock, and that rising tides have forced the Park Service to raise the roads 24 inches. "It's an area that gets submerged," he admits, but it is unlikely that climate change looms large in the Sandy Hook Partners' planning.

Outside the office window, a flock of Atlantic brants, winter residents of New Jersey before their summer flight to the Arctic Circle, were marching around the parade ground. The geese have no reason to fear global warming, or shifting sands either. A wetter, wilder New Jersey will probably be to their liking.

Janine Bloomfield's "Hot Nights in the City" report for Environmental Defense offers a grim scenario for New York in 2100: almost as hot as Houston, swept by floods, wracked by infectious diseases and respiratory distress, and torn asunder at the coastline by erosion and frequent nor'easter storms.

New York has nearly 600 miles of coastline. Four of its five boroughs are located on islands, linked by vulnerable bridges, tunnels, and a subway system that, like the city's three airports, lies less than 10 feet above sea level. Greater New York is the most densely populated region in the United States and a major travel hub; if its airports are even temporarily closed by high water, air travel all over the United States will be disrupted. Low-lying highways that pass through the New York region, including I-95 and I-80, carry much of the nation's truck-based freight. In a sense, then, we are all New Yorkers, and we need to pay close attention to a looming crisis that could affect the city as profoundly as the toppling of its twin towers.

CHAPTER FOUR

# Antigua and Barbuda: Islands under Siege

*Dick Russell*

> Antigua is beautiful. Antigua is too beautiful. Sometimes the beauty of it seems unreal. Sometimes the beauty of it seems as if it were stage sets for a play, for no real sunset could look like that; no real seawater could strike that many shades of blue at once. . . . No real sand on any real shore is that fine or that white (in some places) or that pink (in other places). . . .
>
> —Jamaica Kincaid, *A Small Place*

Fig 4: AgeFotostock/ Doug Scott

Fifteen years after those words were written, I am standing on the shoreline at Runaway Bay. It is mid-January, the height of the tourist season in the Caribbean tropical islands. The sea, as its groundswell rushes in, retains those same wondrous shades—the navy blue, the turquoise, the cobalt. Something is missing: There is no sand. No sand, no tourists. The effect is not so much unreal, as surreal.

Lionel Hurst, since 1988 serving first as ambassador to the United Nations and then as ambassador to the United States for Antigua and the neighboring island of Barbuda, points to the waves crashing against a metal wall just below the concrete patio of the Sunset Cove Hotel. "The owner put in that barrier to try to save his property after Hurricane Luis came ashore in September 1995," Hurst says. "The waves used to break 23 feet further out to sea. But that entire stretch of beach just disappeared, overnight, and it's never returned. So nobody wants to stay here now. The hotel's clients used to walk up a little further and swim. But you can see the same thing has happened in that half-moon area there. Basically, about 1,000 feet of sand has eroded along what used to be one of our most idyllic areas."

Hurricane Luis was the most devastating storm the island had ever seen. With gusts approaching 200 miles an hour and sustained winds of more than 140, Luis damaged 90 percent of Antigua's homes, 65 percent of its business sector, and left seven thousand people unemployed. In a small country now dependent on tourism for 70 percent of its income, virtually all such facilities along the coast needed extensive repairs.

This was the first hurricane to directly strike Antigua since 1950. Yet it was followed, within a week, by Tropical Storm Marilyn. In 1998 came Hurricane Georges. In 1999 came Hurricane José and then Hurricane Lenny. Forecasters called that one "wrong-way Lenny," since it emerged from the west, spinning in the opposite direction of a typical storm, and hammered the westerly side of the island, which usually escapes the brunt.

Clearly, something new was in the wind. "A signal," as Ambassador Hurst puts it, "that something is terribly wrong." Simply stated, warmer ocean temperatures put greater moisture into the atmosphere, two variables that work to power hurricanes. Caribbean-wide, as Hurst would summarize in March 2003 at the World Water

Forum in Japan, storms and hurricanes have risen from an average of 3.5 events per year between 1920 and 1940, to 5.5 events per year between 1944 and 1980, to 13 events per year ever since 1990.

These dramatic weather events have not been evenly dispersed. Because storms in the eastern Atlantic need the earth's rotation to induce the spin factor required to form a cyclone, the further south you are in the Caribbean's "hurricane alley," the less chance of a hit. Barbados, for example, has not had a hurricane since the early 1950s. Trinidad, closest of the islands to South America, is where many of Antigua's yachts have recently relocated. "That's because," according to Hurst, "if you are anywhere between latitude 14 and 22, the insurance companies will no longer provide you with affordable premiums. Against theft and piracy on the high seas there's insurance, yes, but not for hurricanes." Antigua is located at latitude 17 North, the midway point of the Caribbean archipelago. "Even our original airline, Liat," Hurst adds, "is considering moving south out of harm's way."

The average income for an Antiguan is between $10,000 and $12,000 (in U.S. dollars) per year, considered a decent wage in the Caribbean. "But if it weren't for those damned hurricanes, I think we would be doing even better," says Hurst. "The kind of resources we must spend to put humpty-dumpty together again is just a tremendous amount of money we have to borrow. Because we've used up whatever resources we set aside for a rainy day."

We are driving now to Fort James, built by the British in 1739, one of twenty-six battle stations erected during a colonial period that lasted until Antigua and Barbuda gained their independence in 1980. The fort has fallen into disrepair, but from its vantage point you can see the nearby islands of St. Kitt's and Nevis, even Montserrat on a clear day. This morning, a thin layer of ash had been visible on the walkway outside my hotel, emitted by Montserrat's Soufriere volcano.

Hurst relates that shortly before Hurricane Luis hit in 1995, the volcano had erupted full force. It has been continuously active ever since. "We took in more than two-thirds of the Montserrat population, because they had to flee," Hurst says. "Only about 3,000 people are left there. They live in one village essentially, on the north side of the island. Tourism is their sole income these days."

The ambassador is silent for a moment, then continues: "There is some evidence that, when the ocean currents begin to warm up as they have, there is an increase in the amount of volcanic activity. This is just a theory, but that increase could be linked to climate change. Because, you see, colder ocean fronts keep the surface of the earth sufficiently cool that movement of molten rock on the surface is kept to a minimum." (In July 2003, after my visit, Soufriere erupted again, sending volcanic ash 40,000 feet in the air and spewing rocks and mud down on houses, but causing no injuries.)

More hurricanes and volcanic eruptions are just the most extreme manifestations of what is happening to the leeward islands of the West Indies. Make no mistake: Antigua is still achingly beautiful. Including day-trippers from the cruise ships, more than half a million visitors continue to arrive annually to vacation at places such as the Jolly Beach Resort, enjoying wintertime temperatures in the 80s and cool rum punches before an evening feast of fresh grouper against a backdrop of Caribbean steel drums. But there is trouble in this Westerner's paradise, and local Antiguans sense it all around them in myriad ways.

## A Changing Island

Junior Prosper teaches geography in a secondary school. He also does volunteer work for the island's Environmental Awareness Group. On a Sunday morning, with his wife and daughter in the back of his jeep, Prosper is taking me to parts of Antigua that I might not otherwise see. The island is only 9 miles long by 12 miles wide, 108 square miles in all. Of its 73,000 population, some 30,000 people live in the capital of St. John's. But we are bound for the hinterlands, past Dark Wood Beach and Crab Hill.

I am grateful Prosper is at the wheel. Not only because I cannot get used to driving on the left side of the road. Or because everyone else seems to know just where all the potholes are, as they barrel around corners somehow avoiding goats, cows, and riders on horseback. But also because there are no street names, or even any signs announcing what little town you are now passing through. My mind occupied by all that, I would certainly never have observed solo what Prosper is showing me: Antigua's three distinct geological areas, as the landscape shifts from volcanic to clay to limestone.

As he drives, Prosper points out the tall royal palms, planted strategically since colonial times to follow streamways. "Once you pick up a royal palm, you know it would lead you down into some settlement." He shows me the clamonberry trees, which provide shade to the streams and control evaporation. "If you talk to the locals, they call it turtleberry tree, because you find that particular bird—the turtle dove—nesting there. Areas on Antigua where they have cleared these trees, the whole pond system has dried up."

Prosper has been detecting other differences, unnatural differences, in this landscape. "The overall vegetation of Antigua has changed in the last 20 years. You tend to be seeing more empty spaces." I recall Hurst telling me that, in an already dry climate where average rainfall used to be 39 inches a year, it is now down to 24 or 25 inches. That's despite the increase in hurricanes. "What I notice, too," Prosper continues, "with the large tree crops—coconuts, mangoes, breadfruit—the time of year that they bear and send fruit has changed. The size of them, and the amount of juice in them, has also changed. They're not as plentiful, and the breadfruit for example is half the size it used to be. The thing is, you're not sure whether this has occurred because of the hurricanes coming through so rapidly, so the trees don't get a chance to regenerate to a point of maturity. But it is definitely hotter, and drier. So there is less moisture in the ground, impacting on the size of the fruit."

We have reached the Wallings Forest, the largest remaining tract of moist evergreens that covered Antigua before the Europeans started clearing three centuries ago. Migrating warblers and over a half-dozen bird species found only in the Caribbean reside here, in trees as high as 80 feet. We park and hike uphill to the Wallings reservoir. "Four channels made out of concrete that bring the water to this dam are all blocked up, with debris from the hurricanes," Prosper informs me. He also recalls coming here in 1977, with some of his first students. They brought along a thermometer. "We measured the temperature here at a normal mid-day at a cool 78 degrees. Today, it's roasting."

Beyond lovely Willoughby Bay, where famed guitarist Eric Clapton has built an upscale drug rehabilitation center, we enter a region of red hills. The roadside is ringed by pipe organ cactus, agave, and a ubiquitous thorny shrub known here as "cassie." It has sprung up where sugar cane plantations once predominated—more than 160 of them at one time, abandoned after a gradual price decline forced the colonial owners out of business. Today, the old stone grinding mills dot the island at intervals, resembling abandoned watchtowers.

Our next destination is Potworks Dam, primary source of drinking water for Antigua. A scarcity of high hills and forest growth, as well as its position relative to the equatorial rain belt, leaves the island with no rivers and thus frequently subject to drought even before the advent of climate change. In 1984, Prosper remembers, the situation got so bad that freshwater had to be imported on barges from Montserrat and Dominica. Shortly after that, a desalination plant was built to help take up the slack. A layer of clay soil a couple of feet down at Potworks keeps the water naturally near the surface but, with more water also being pumped out to irrigate nearby farms—and particularly as a result of the decline in rainfall—the dam is well below capacity.

Brian Challenger, who oversees the Antiguan government's nascent climate change project, described the situation like this: "2001 was our driest year on record and even that oversimplifies it, because when you combine this with another dry year in 2000, the soil moisture level is extremely, extremely low." The moment the island was not being slammed by annual hurricanes, in other words, the general decline in rainfall was amplified. "This affects your biodiversity, which affects your watersheds, which affects your touristic appeal," Challenger summarized. "When the groundwater or surface supply gets used up, we have to rely solely on desalination—which, of course, is much more costly, and passed on by the utilities to the consumers."

The technical director for the Ministry of Planning, Daven Joseph, elaborated that a rainy season which used to start in July has not been arriving until October for the past several years. "Another thing we are noticing," Joseph added, "is a strong northeasterly [wind] we would normally have from December into February has reduced significantly. So the weather is warmer in those months." A warming trend is especially noticeable, Challenger told me, in terms of higher nighttime temperatures, year-round.

These are the subtle shifts, the things only a permanent resident would see and feel. At the same time, these alterations in nature's patterns are accompanied by development schemes that exacerbate their impacts—and also cloud the picture of what, or who, is really to blame. The effects are synergistic. Circling into the driveway of a hotel whose perimeter of mangrove trees has recently been dredged for a prospective golf course, Prosper explains: "The developers are supposed to stay 100 feet away from the shoreline, but people are quarreling over where the shoreline actually starts. There have also been problems with expansion of many hotels that have come to the coastline, when large stones are brought in to make barriers, piers, and so on. Well, now you don't have to wait for a hurricane. Anytime there's any severe weather, or even the pass of an easterly wave, the waves crest [over the barriers] and water gets into the hotels."

The island's original name was Wadadli (a distinction now reserved for the local beer) or "well-water island." Later it was renamed Antigua or "anti-agua." One wonders at the prescience of whoever conjured up the latter name.

## The Mangroves Are Gone

The ideal vantage point for exploring what is happening to Antigua is from the sea. On a Friday morning, along with a half dozen other visitors, I embarked on "Eli's Eco Tour," a day-long voyage around the island from the Caribbean Sea on the west side to the Atlantic Ocean on the east. It is skippered by 30-year-old Eli Fuller on his 34-foot, center-console motorboat, the *Isis*. Fuller was born and raised on Antigua. His grandfather, who arrived in 1941 as America's vice-consul, stayed on to open the island's first beach hotel. Eli's father is a prominent local attorney.

We head out to sea from Jolly Harbor, with its 150-slip marina, resort, and golf course, past what Fuller says used to be a "very healthy" mangrove swamp. Woody plants that have adapted to living in salty, muddy waters, mangroves can grow to be trees as tall as 25 feet or remain as scrubby growth in drier areas. They serve as spawning grounds for many species of fish, absorb flood waters, and reduce the impact of ocean waves. Here a resort owner, receiving complaints about mosquitoes, obtained government permission to dredge the harbor and construct a seawall (along with, eventually, five hundred villas). "Because of removal of the mangroves, runoff from the hills sends more sediment into the sea water," Fuller notes. Hurricane Luis exacerbated the situation, delivering a massive pool of mud that filled in the bay. Its waters have never been as clear since.

"Watch out for your hats," Fuller says as we near Reads Point. "These rolling swells, I'm gonna slow down, but expect a little drop as we go over them." Turning north up the Caribbean coast, we hurtle past Antigua's lone nude beach ("the only place the police turn a blind eye to naturism"), pick up a few more passengers in St. John's, and continue on past Runaway Beach. The erosion I had witnessed standing on the shoreline with Ambassador Hurst is even more startling when viewed from the sea.

Fuller shakes his head. Just south of where we had started out, he says "the severe erosion isn't just because of the sea and the bigger groundswells we've been seeing in the wintertime. After the '95 hurricane pushed all the sand over into the swamp at Dark Wood Beach, when it was still uninhabited, trucks took the sand away in a government-sponsored mining operation. Then, at Pinchin Beach, the same thing happened; this time, a local family who owns the land there mined all the sand from behind the beach. This is when I wrote to the Prime Minister, through Sir Ronald Saunders, to put a stop to it. It did stop for a while pending an 'investigation.' As usual, nothing happened and sand was later mined quietly from the same beautiful beach."

Our cruise continues, with Fuller pointing out yachts that charter for $20,000 a day, a resort where the rooms go for $2,400 a night, and another hotel "whose gimmick is, they can have all 350 of their guests on the water at the same time" on various types of sailing craft. The new sport of kite surfing has recently made its debut. "Using a board the size of a snowboard, essentially you're sailing backwards and forwards using a kite to pull you," Fuller explains. "Depending on how you turn the kite, once you're good it can lift you 25 or 30 feet in the air, and you're doing all kinds of flips and spins while you're up there."

Fuller points out a barrier reef where the waves are breaking, which continues for about 5 miles up the coast. In one of the recent hurricanes, though, the reef was badly battered. In certain sections where it came almost to the surface, the reef dropped 5 feet. This has resulted in more ocean swells passing over it than before, and thus an increase in coastal erosion. Coral is a very delicate ecological structure, and Antigua is already seeing 50 percent mortality among its reef systems. The higher surface sea temperature rises, the greater the "bleaching" of coral. According to a May 2001 report by the United Nations Development Programme's (UNDP) Climate Change Project, coral reefs around Antigua and Barbuda are "currently growing precariously close to their maximum temperature tolerance of 30° C [86° F]."

Nor is global warming the only factor. As we pause to go snorkeling near the offshore Guiana Island, Fuller tells us about the parrotfish. "They're a very popular fish for cooking here. People deep-fry them with lots of seasoning. You'll see spear fishermen with 20 or 30 of them on a string. But parrotfish eat the algae on the coral, preventing the algae from taking over and wiping out the coral. So if you fish out thousands of parrotfish, which is happening, your reef is severely affected." This is coupled with the disappearance of another algae-eater, the spiny sea urchin, that suffered a mysterious mass die-off in 1985.

Antigua's waters are prolific with other species of fish, but its fishermen are also experiencing the effects of climate change. During Hurricane Luis, about 16 percent of the island's fishing fleet was destroyed or lost, another 18 percent damaged. The fishery is impacted, too, by alterations to habitat, where juvenile fish customarily thrive in mangroves and sea grass beds. The distribution of the latter is controlled by water depth and salinity. Due to anticipated sea-level rise, the UNDP report predicts, "There would be little or no mangroves in Antigua by 2075, since the coastal slopes of most areas do not allow for landward retreat." The study says the condition could be reached as early as 2030.

Sea levels across the Caribbean region are expected to rise between 11.8 and 19.7 inches over the next 50 years, higher than the predicted global average of 1 foot. As we go ashore for lunch on Great Bird Island—home to the planet's rarest snake, the Antiguan racer—I ask Fuller about what he has noticed over the years. "I think the water level is already rising, more than what other people are saying. I hear talk about a centimeter every certain number of years, but it seems as though it's more dramatic than that. You can see along here, we used to have much more sand. Now it's low tide, and the water is fairly close to the edge there. It's not just that sand has been eroded, but that the water level has risen quite a bit. That's the same for a lot of places around Antigua."

I look around the shoreline, at the sea grape trees with their big round yellow leaves that leave a pungent sweet-and-sour flavor in your mouth. At the century plants with the yellow and white flowers blooming at the top. At the bushy, land-growing black mangroves. At the wild sage you mix into "fish water" in cooking your fresh-caught red snapper. I look, and I wonder, wistfully, what a visitor to Great Bird Island will see in 50 years.

On our trip back, beyond Guiana Island and its surfacing green turtles, beyond a white egret diving for crabs over a tidal pool, we pass near the original hotel that Fuller's grandfather constructed in 1950. "They closed after Hurricane Luis hit in '95," Fuller says, "and rebuilt just in time for Hurricane Georges in '98. The hotel got wrecked again, but my cousin spent a lot of money fixing it up. Last year it reopened."

Fuller is silent for a time, steering us past exclusive Long Island where England's Lord Sainsbury, English author Ken Follett, and television's Robin Leach from *Lifestyles of the Rich and Famous* have their homes. Then Fuller offers what he has been quietly musing about. "We had a hurricane in November as well. That's another thing that never happened before, hurricanes that late. That was Lenny in '99. Luckily, it died down just before it got to us. So we didn't have a lot of wind, but we had 23 inches of rain in 24 hours, the highest rainfall we ever recorded, causing landslides all over Antigua. On the main island over here, big pieces of hillside just dropped off."

## Barbuda: Half an Island?

The isle of Barbuda is 27 miles north of Antigua. There is evidence the two were joined in recent geological times, but today in some ways Barbuda could be on a different planet. Although about two-thirds the size of its sister island, Barbuda has a mere fourteen hundred inhabitants, most of whom live in the town of Codrington. The town is named after Christopher Codrington, a colonial governor of the Leewards who once signed a 200-year lease with the British government for Barbuda—with the price limited to "one fat sheep." The island is almost completely surrounded by reefs, on which more than two hundred shipwrecks have occurred. It is also very flat. So flat that, as Daven Joseph put it, "If the sea-level rise prediction comes true, we are going to lose more than half of Barbuda in the next 50 to 60 years."

As a limestone island, Barbuda has extensive underground water supplies. But here is what the UNDP's Climate Change Project has to say about the situation: "Most of the main aquifers, wells, sinkholes and water bodies in Barbuda are located relatively close to the coast. The depth of the water table is generally less than 1.5 meters in the lowlands. Any increase in sea level can affect the level and salinity of the groundwater supplies. With the projected sea-level rise, the main aquifers and wells may be fully or partially inundated and the groundwater supplies could become permanently lost. This will threaten the entire economy of the Island of Barbuda."

There are two propeller-driven flights a day from Antigua onto Barbuda's little airstrip. Upon a morning arrival, I phone a fellow who rents four-wheel-drive vehicles for $50 a day out of his home. Getting around Barbuda is challenging. None of the streets, or buildings, are identified with names or numbers. The island's main road poses an obstacle course of tremendous potholes, along with wild donkeys and wild goats. The gentleman who has agreed to show me around, I am told, lives in the only residence on a bumpy dirt road to the left at the local hospital.

John Mussington, wearing glasses and a baseball cap, emerges from his modest white wood-frame house with a welcoming smile. He is in his late 30s and has five children. Principal of the 350-student Holy Trinity School, Mussington holds a degree in marine biology from the University of the West Indies, and is also the Environmental Awareness Group's representative on Barbuda.

The first thing Mussington tells me about is his house, which he finished constructing in 1994, the year before Hurricane Luis struck.

"Fortunately, I built it with the intention it would withstand a hurricane," he says. He points to the concrete blocks that elevate the house several feet off the ground, tied to a concrete foundation of equal depth. Although the Codrington Lagoon is not visible through the trees, Mussington lives less than 1,000 feet away from its eastern shoreline. "During the height of the storm, just after the eye passed, when I came out on my verandah I was totally surrounded by seawater," he recalls. "About six inches higher, it would have been flowing through the house." While the water there drained quickly once the storm lifted, most of the other homes in proximity of the shallow, 7-mile-long lagoon had already been devastated.

Leaving Mussington's house, we head south for a few miles. Then he instructs me to turn off the main road onto a narrow, increasingly sandy track bordered by low, desert scrub. As the track finally becomes impassable, we begin to hike toward Palmetto Point. Here, for centuries, massive sand dunes as high as 50 feet loomed just inland from the Caribbean. They served as nature's protective barrier against saltwater intrusion into the freshwater aquifer. They enabled beans, peas, watermelon, and corn to flourish in farmer's fields. While Mussington is telling me this, I search the horizon in vain for any semblance of a sand dune.

He explains: "They began mining the sand in the mid-1970s. Local politicians. It became a multimillion-dollar industry. The sand was exported to Antigua, as far as Puerto Rico, the Virgin Islands, even down to St. Lucia. It's used in construction and things like that. Everything has basically been taken out. So this flatland you're seeing is actually sitting on the water table right now. During the last hurricane, Lenny, there were still some ridges that the storm surge would encounter. Not anymore. So the danger that's facing this area is in the event of another hurricane or severe storm. If you get, say, a 10 to 15 foot surge, it's going to reach the narrow connection between Palmetto Point and the lagoon, which is only 200 meters [650 feet] wide. Once the saltwater floods into that area, out goes the agriculture."

Returning to the vehicle, we turn around and take another branch road until it, too, reaches a virtual dead-end at a small hill. From here is the best view of Palmetto Point, and a 14-mile swath of white sand beach that is touted as one of the most pristine such stretches left on the planet. "We have some unique vegetation growing on this point here," Mussington continues, "and we're already losing it to the Caribbean. In some cases we don't even know what we're losing, because many of the plants have not been catalogued yet. This, for example, is called torchwood. It has so much resin in it, that it's believed the Indians who used to live here would tie a pile of this wood together and light it to form a torch. We also have mauby. The bark can be used to make a drink that has medicinal qualities as a diuretic. Anything the doctor would prescribe to allow your body to lose more water, this would do naturally, as well as increasing appetite. But as sea level continues to rise, we will lose the torchwood, the mauby, all of this."

As on Antigua, Mussington has witnessed a steady decline in the health of Barbuda's coral reefs. "When I started working here in 1983, you had a lot of living coral, fringing the western side and especially on the northern side of the island. Today, those same reefs are almost totally dead." He ticks off the same reasons I had heard from Antiguans: overfishing of the parrotfish and die-off of the spiny sea urchins that grazed on the algae, combined with damage from hurricanes and "bleaching" from warmer water temperatures.

It is the reef system, along with the lagoon, that has provided Barbuda with its main industry—a historically thriving lobster fishery. The island's lobsters are exported to Antigua, Martinique, Guadalupe, and other spots in the Caribbean. They used to weigh as much as 12 pounds. "Now you don't get many of those," Mussington says, "because as with everything else, the fishery has been decreasing."

After a drive into the highlands (at 100 feet, Barbuda's apex), Mussington is anxious to show me the Codrington Lagoon Bird Sanctuary, where the largest colony of Magnificent Frigatebirds in the Western Hemisphere comes during breeding season. Black birds with tiny feet and chicken-sized bodies, but with massive wing-spans of up to 8 feet, these have been termed nature's most perfect flying machine. Mussington recalls doing a count in 1998 that detected as many as five thousand nesting pairs.

It is indeed stirring to watch a frigatebird soaring effortlessly above a pond, then swooping suddenly into the water to spear a fish. But do I detect a plaintive sound in the male's warbling cry?

"When you sit here totally surrounded and dominated by the sea," Mussington says, "and that is your life and your livelihood, global warming becomes a big deal."

I ask whether there is any talk of utilizing renewable energy on the island. "It's a minority, but I'm trying to push for it as much as possible," he responds. "Because we are flat, the solar intensity you get here is higher than in most of the Caribbean islands. The angle at which the rays of the sun hit us, and the duration, makes us an ideal place for solar technology. Also, with the prevailing winds we have, setting up wind towers could meet the needs of 1,400 people comfortably and pay for itself within a matter of 10 years. These things are expensive in terms of initial investment, but in the long run we are isolated and our best bet is to try to go in that direction. For example, when Hurricane Luis passed through, this island was cut off because neither airplanes nor boats could get here for a couple of weeks. In that circumstance our fuel needs started to be used up, which meant our power generation plant could not operate."

In terms of future development for tourism, Mussington is adamant that setback limits must be implemented. "We cannot continue to have investors coming and building hotels right on the beach!" He suggests I drive down-island toward Coco Point, and try to sneak a look behind the gates of the exclusive K-Club.

So, after dropping Mussington home and sampling the Palm Tree Restaurant's spicy bull-foot soup, stewed goat, and pigeon peas rice (all quite delectable), I wend my way south again, keeping an eye out for wild boar that are occasionally sighted along the landscape. The main gate to the Italian-owned K-Club, where beach villas rent for almost $3,000 a day, is barred. The club is situated in a 230-acre park with a nine-hole golf course and two tennis courts. I drive on below the white bungalows, on a mile-and-a-half-long frontage road. Finally I park outside a rear entryway and simply walk through some trees onto the beach.

It is astounding how close the K-Club's villas are to the ocean. Today's calm, limpid sea laps up to within about 30 feet of the dwellings. A few people are out getting a tan, but for the most part the resort appears almost empty. At last I enter the central clubhouse with its advertised 146 columns and a wine list to please royalty. The manager is out, so I content myself chatting with the young man in charge of the boathouse. He's pleased to offer names of some of the recent guests: "that *60 Minutes* guy," Ed Bradley, was here over Christmas. Sylvester Stallone dropped in once "on a big boat," Princess Diana had stayed here with her kids.

The young man was working at the club in 1995 when Hurricane Luis descended. "Lot of high swells and wind took some sand inside the villas. Basically Luis ripped the hotel apart. But they rebuilt it."

And had anything changed when the K-Club was rebuilt? No, the young man says, it was constructed "exactly the same."

## Their Hands Are Tied

Inside the Ministry of Planning in the Antiguan capital of St. John's, Daven Joseph leans across his desk and talks of the need to "train our people in coastal zone management and development." Removal of beach sand has been restricted but, he adds, "It is not enforced properly because we don't have the capacity to patrol the beaches." There has been, he admits, no effort to replant mangroves, "but I think that can be one of the more significant programs."

At another cramped office not far from the airport, Brian Challenger notes that summer planting of mangroves is being undertaken by fishermen and the environment department, "but it certainly doesn't equal what has been cleared." The Ministry of Public Utilities, where he works, is preparing a study on climate change and tourism that will contain other suggestions such as setbacks for beach hotels and elevations of 2 feet rather than 1 foot for buildings.

Challenger is also examining whether health problems could be exacerbated by climate change. "For example, the leading cause of death in Antigua and Barbuda is cardiovascular. If you have to withstand more heat, your heart must work harder." The Caribbean Epidemiology Center has simultaneously undertaken a 3-year project examining what warmer days will mean for the breeding of dengue fever and malaria, diseases already endemic to the region. The last major dengue outbreak in Antigua occurred in 1981, with seventy-seven identified cases. Temperature rise will likely bring more *Aedes aegypti* mosquitoes as carriers.

A new study published in the journal *Science* in July 2003, based on data that British scientists compiled from 263 separate coral reef sites in the Caribbean, painted a grim picture of their health: Over the past three decades, the amount of coral cover has dropped 80 percent. As long ago as 1990, Jamaican authorities estimated it would take $462 million (in U.S. dollars) to protect coastal tourism on its island alone. According to Clifford Mahlung of Jamaica's National Meteorological Service, "Climate change is likely to bring declining soil fertility, droughts, flooding and deforestation, and the displacement of many who depend on coastal resources." Mahlung cited amplified problems with the disposal of sewage and solid waste. At the country's famous white-sand beach of Negril, 33 feet washed away between 1995 and 1998. There was a 60 percent reduction of rainfall in one Jamaican parish, and some hillside temperatures are up an alarming 7.2° F.

With climate change as its primary focus, the Caribbean islands have come together with their counterparts from all oceanic regions of the world to form an Alliance of Small Island States, established in 1989 and now having forty-three members and observers who meet regularly at the United Nations. The 2003 World Water Forum held in Kyoto, Japan, resulted in formation of a Joint Caribbean-Pacific Program for Action on Water and Climate. Particularly noted was the high water demand needed to sustain tourism (in Jamaica, e.g., the tourist sector demands 10 times more water per capita than the domestic sector).

Still, as Challenger said during our interview: "We will do what we can, in terms of improved efficiencies, but globally it's a drop in the bucket. When you look at the amount of greenhouse gases that we emit, it is less than a small town in the U.S. Even less than probably one *factory* in the U.S. And if Antigua and Barbuda should set a standard of 50 miles per gallon for our vehicles, it means nothing. We are dependent on Western standards, because that's the reality of the world. So if the U.S. continues to finesse us up, whatever we do in the long term is spinning a top in the mud. We'll be trying to do things that are futile."

Perhaps no one has summarized the situation better than Ambassador Hurst. Speaking to the World Water Forum, he said: "The most populous and wealthiest of the world face a moral challenge greater than colonialism or slavery. They are failing in that challenge. Men have lost reason in the fossil fuel economy. Materialism and self-interest are winning out over social justice. Developed countries have been conducting a dangerous experiment. The skies do not belong to the countries of North America or Europe—they are for all of us. Inhabitants of small islands have not agreed that their tomorrows will be the sacrificial lambs on the altar of the wealth of the rich. The U.S. has deliberately evaded the Kyoto Protocol. A few important world leaders are in denial. The time has come to put civilization on a new trajectory."

## Chapter Five

# Asia: Clouds Got in the Way

*Jim Motavalli*

The Indian coastal city of Mumbai, formerly Bombay, is home to India's vibrant film industry ("Bollywood") and probably boasts more cell phones per capita than any other city on the subcontinent. But it is also home to one of Asia's largest slums. Half of Mumbai's population lacks running water or electricity, and the smoke from hundreds of thousands of open cooking fires joins with the sooty smoke from two-stroke auto rickshaws, belching taxis, diesel buses, and coal-fired

Fig 5: AP Photo/Nick Ut

power plants in a symphony of air pollutants. Breathing Mumbai's inversion-trapped air, they say, is the equivalent of smoking twenty cigarettes a day.

India's capital city, New Delhi, is even worse than Mumbai. Tourists get a full measure of it when they travel by teeming road or slow train to the tourist mecca of Agra. Just south of New Delhi, they encounter a hell's circle of industrial towns, including Faridabad and Ballabgarh. The *Hindustan Times* has called Faridabad "a gas chamber," where "citizens are dying a slow death due to the high level of pollution."

Faridabad has fifteen thousand three-wheel diesel vehicles known as auto rickshaws, many of them refugees from Delhi, which now allows only rickshaws powered by compressed natural gas. After the rickshaws grew out of control in 2002, they were ordered to operate only on alternate days. Faridabad also has eighty dyeing and printing industries, which pollute rivers and groundwater. There are also two hundred electroplaters, whose by-products include contaminated waste water, cyanide, and hydrochloric acid. In 2001, 45,000 people sought treatment for respiratory problems in Faridabad, says Dr. S. P. Singh Bhatia, the region's chief medical officer.

According to a 1997 air quality survey by India's Central Pollution Control Board, sixty-nine of India's seventy principal cities are moderately, highly, or critically polluted year-round. The locally popular term for the hazy clouds that blanket these places, causing sore throats and aggravating allergies, is "fog." Journalists who stay at the "five-star" Taj View Hotel in Agra are well acquainted with the fog because it completely obscures the view of this most famous of landmarks. Tourists, 1.8 million of them annually, now reach the Taj Mahal in a small fleet of electric vehicles, a belated and probably futile attempt to preserve Shah Jahan's seventeenth-century palace from the eviscerating effects of pollution.

The Taj Mahal is no longer the dazzling white of tourist brochures. Although it still looks, in the words of a nineteenth-century British surveyor general, to be made from "pearl or of moonlight," it now has a distinctly ivory cast, shading into yellow. The Indian Conservation Institute claims that the yellowing is simply a build-up of dirt and the building needs to be cleaned. The main culprit, however, is clearly airborne sulfur, produced by some 150 iron foundries, auto and motorcycle exhaust that the electric vehicles do little to alleviate, and the huge Mathura oil refinery 26 miles away. (The latter was built despite a parliamentary committee's conclusion in 1979 that it was "the worst possible [site] . . . from the archaeological, ecological and environmental points of view.")

Sulfur, emitted at ten times India's standard, combines with oxygen and moisture in the air to produce sulfuric acid, which attacks and discolors the Taj's marble. Pieces of marble have begun flaking off the priceless building, and inlays have come loose. The Taj is still unquestionably a magnificent queen, but it is starting to look just the tiniest bit dowdy.

Mahesh Chandra Mehta, a prominent environmental lawyer in India, filed a suit with India's Supreme Court in 1984, claiming that pollution was ruining not only the Taj but also the health of the people of Agra. Twelve years later, in 1996, the court finally ruled in favor of Mehta's suit. Coal-based brick kilns were ordered shut down. The biggest factory in Agra, Sterling Machine Tools, switched from coal to natural gas. In all, some 292 coal-based industries were asked to switch to gas fuels or close down by the spring of 1997. (Mehta won the Goldman Environmental Prize in 1996 for his work protecting the Taj.)

"The general cleanliness of Agra city has been ordered," read one local account, but actual progress was slow. Many small companies lacked the funds to make the switch to natural gas. The basic cost, according to UNESCO, was $75,000 to $100,000—a quarter of annual sales for some of these operations.

In 1999, India's Supreme Court again ordered recalcitrant industries to clean up. Meanwhile, Agra's Iron Founders' Association has been fighting back, claiming that efficient natural gas technology is not yet ready and that shutting down the foundries will idle thirty thousand workers. Faced with closure, some factory workers burned their bosses in effigy.

By 2000, the city moved cars and some small shops away from the immediate vicinity of the Taj, but few factories were shut down. That same year, then U.S. President Bill Clinton visited Agra to sign an environmental agreement providing $45 million for energy-efficiency programs in India. "Pollution," he said, "has managed to do what 350 years of wars, invasions and natural disasters have failed to do. It has begun to mar the magnificent walls of the Taj Mahal." Clinton decried the "marble cancer" that had begun to erode the building. "I can't help wondering that if a stone can get cancer, what kind of damage can this pollution do to children," he said.

There is a general sense in India that the "fog" is a natural phenomenon, an impression mirrored in India's chauvinistic press. In the *Indian Express,* for example, Dr. J. G. Negi of the National Geophysical Research Institute opined recently that India "has nothing to worry about" when it comes to global warming, since India does not produce much greenhouse gas and, anyway, the higher levels of carbon dioxide will lead to "increased vegetation." Instead of a warmer planet, Dr. Negi thinks we will be in the grip of a new Ice Age by 2010.

In fact, however, India is the world's fifth-largest producer of global warming gas (the United States is first), and emissions there are bound to get worse as the population soars past one billion and private car sales (up 58 percent between 1998 and 1999) skyrocket.

Early signs that the climate is already changing are abundant. In 1999, a heat wave killed hundreds of people and led to thousands of new cases of gastroenteritis and cholera in New Delhi. Major droughts hit the eastern Orissa state in both 1999 and 1998, the latter resulting in two thousand deaths. The heat wave continued into 2000, affecting some fifty million people. Record temperatures dried up wells, rivers, and streams and resulted in water crises in eleven of India's thirty-one states. In hard-hit Orissa, temperatures reached 118° F. Government offices closed in the afternoon, trains carrying drinking water were mobbed, and prisoners rioted, shouting, "Give us water or kill us!"

In 2002, the country was devastated by large floods that killed thousands of people in eastern states, and renewed heat waves and drought in much of the country. The India Meteorological Department described 2002 as "the first all-India drought year" since 1987. One of the wettest places on earth, the Khasi Hills of northeast India, once set a record with a rainfall of 1,000 inches in a year. But by 2003 it was importing truckloads of water from the plains. The hill region of Cherrapunji, which has been heavily deforested, received 363 inches of rain in 2001—about what it got in just 1 month of 1861.

The drought caused major crop loss and starvation, killing ten children in Rajasthan alone. Of forty-five districts in Madhya Pradesh, only three had normal rainfall and seven had no rain at all. In Uttar Pradesh, twenty-one districts were severely affected. Singers toured parched villages and offered songs in praise of the Hindu rain gods. "This heat wave is a signal to global warming," warns M. Lal of the Centre for Atmospheric Sciences. "India is simply following the global trend of higher temperatures." Lal predicted that India could heat up dramatically by the end of the century, with "a major impact on the country's food production and water reserves."

## The Climate Summit

For much of India, the heat is already on. In late 2002, New Delhi hosted a United Nations conference on climate change, attracting bureaucrats and political leaders from around the world. Left out of the air-conditioned event, however, were the voices of average Indians already affected by global warming. Those people were heard at an altogether different but simultaneous forum, the Climate Justice Summit, organized by the Indian Climate Justice Forum.

"It is the poor and the marginalized who are, and will continue to be, the hardest hit by the impacts of climate change," says Rita Nahata of the Forum. "The biggest injustice is that the hardest-hit communities are the least responsible for creating the problem." As if to illustrate her words, the Forum displayed such low-impact tools as a bicycle-driven water pump and a hand-operated food processor.

The Summit was quite a bit more boisterous than the United Nations conference taking place just a few miles away, and included people chanting "water, land, forests are ours" and "multinationals go home," punctuated by the rhythms of a drumming contingent.

The grassroots event attracted an estimated five thousand people, including labor organizers, farmers, fish workers, one hundred rickshaw drivers, indigenous people, and street children. The latter, organized by the Butterflies—a kids' union that offers schooling and a credit union—paraded with papier-mâché puppets that depicted the United States eating the world.

If the mood was sometimes festive, the purpose was deadly serious. The Summit adapted a declaration, dated October 28, 2002, and delivered to the United Nations meeting at the close of the rally, that declared climate change to be "a matter of life and death for most communities in India." The impact of global warming, it said, is "disproportionately felt by the poor, women, youth, coastal peoples, indigenous peoples, fisherfolk, dalits [sub-caste 'untouchables,' of whom there are 240 million in India], farmers and the elderly."

The protesters announced their opposition to "market-based mechanisms and technological 'fixes'" that "only exacerbate the problem," and blamed "unsustainable production and consumption habits" that "exist primarily in the North, but also among elites within the South."

Climate change is far from an abstract concept for many Indians. In "Fishing in Troubled Waters," an article published by the online Corpwatch.com, Lalitha Sridhar describes how India's fishing communities—well represented at the Summit—are especially hard-hit by global warming.

In Tamil Nadu, India's densely populated southernmost state on the Indian Ocean, fishermen whose ancestors fished these waters for millennia are suddenly idle. "I hate my life," says Shekhar, a 40-year-old fisherman. "It is not like my grandfather's time. Nothing is the same anymore—even the fish are gone."

According to a study by Dr. Herman Cesar of the Free University in the Netherlands, surface water temperatures in the Indian Ocean were often 7 to more than 10° F above normal during a 5-month study period in 1998. During the same period, which coincided with the El Niño southern oscillation (and exacerbated by global warming), fishing catches dropped significantly. "Fishmeal production fell by 10 million tons—about 10 percent of the global fish catch—and entire species such as horse mackeral, mackeral and hake were acutely scarce," Sridhar reports.

Dr. Cesar adds that coral bleaching, which has begun to affect even remote and previously pristine atolls in the central Indian Ocean, will further aggravate the fishing problem. Global warming has also devastated the mangroves that act as a nursery for marine diversity in the southern Gulf of Mannar, home to a marine biosphere reserve.

Activist T. S. S. Mani, who works with the fishers of the Nochchikuppam commune, says that climate change has "a direct and terrible effect on the livelihood of coastal fishworkers. They rate amongst the poorest of the poor but their concerns are completely marginalized. . . . Although they may not understand global warming and greenhouse emissions, their traditional knowledge of the oceans, fine-tuned over centuries, should be providing key insights. But artisanal fishworkers have become victims of development instead of being participants in the process."

## The Asian Cloud

The global warming effects that are pushing Indian fishermen to the brink of starvation are part of a larger picture. American military pilots flying over the Indian Ocean from the U.S. Air Force base on Diego Garcia first detected the presence of a large, dense cloud of sooty pollution over Asia in the 1980s. Since then, it has regularly appeared in satellite photographs and been tracked by research ships.

In 1999, a team of scientists funded by the National Science Foundation began a $25 million surveillance of the Indian Ocean and discovered that the huge layer of haze covers 10 million square miles, approximately the size of the continental United States. The scientists were astounded by the size of the cloud, which is made of tiny sun-blocking pollutant particles called aerosols.

Not only was the enormity of the pollution an issue, but its composition as well. Here was evidence that soot, in gigantic concentrations, could influence climate almost as much as carbon dioxide. The haze, at an elevation of 1 to 2 miles, covers most of the northern Indian Ocean, the Arabian Sea west of Mumbai, and the Bay of Bengal. The source: air pollution from India and China, produced by hundreds of millions of cooking fires and coal furnaces, blown out to sea during the winter monsoon season.

We are not talking about a hypothetical effect here. The Asian cloud has been integrated into the daily life of a region that is home to 60 percent of the world's six billion people. For John Hayes, a British resident of Bangkok, the ever-present choking pollution finally forced him to uproot his family and return to England. In a *Guardian* story entitled "Life under the Asian Brown Cloud," Hayes wrote about the "two-stroke motorcycles spluttering burnt oil, 10-wheel trucks converting inferior, adulterated and cheap diesel into dense black clouds, and nice shiny imported limousines spewing out more fumes." The journey to Hayes's children's playground, only a mile away, could take an hour of lurching through traffic and smoke. The kids, he says, visited the hospital "at least once a month with allergies that caused irritation to the eyes and nose, or with the vicious effects of a tropical common cold. This usually meant not just a runny nose, but a high fever accompanied by aches and pains. The sense of helplessness that any parent experiences when their child is sick is exacerbated by the severity of the symptoms."

Writing from Bangkok for Melbourne, Australia's *The Age*, Alex Spillius reports that a dinner party in Asia "almost invariably produces a round of pollution stories: the friends whose children developed chronic asthma in Delhi; the horrors of heavy industrial waste in Beijing; the filth in the sea off the beaches near Bangkok. But anybody who endured the 1997 haze in Kuala Lumpur or Singapore generally wins the raconteur's first prize."

Spillius further states that "the smog caused by forest and plantation fires in Indonesia was so intense that visibility was reduced to meters, schools were closed and children ordered to stay indoors. The populations of whole cities wore protection masks . . . And still the fires continue to blight Southeast Asia each year." In early 2000, some twelve hundred forest fires were detected in Sumatra and Borneo, sending pollution levels soaring in Indonesia and neighboring Singapore. The fire-borne haze became worse in 2001, when residents of Singapore and Malaysia were forced to don respirators. In central Kuala Lumpur, smoke obscured the hills around the capital. In southern Thailand, residents of five provinces were urged to stay inside or wear face masks.

The Asian cloud has spurred a run on the sales of air conditioners and air purifiers in Malaysia, and led to a new cocktail known as "the haze." Air and auto crashes were blamed on the poor visibility under the cloud.

Dr. Veerabhadran Ramanathan, a professor of atmospheric and ocean sciences at the Scripps Center for Clouds, Chemistry and Climate, is co-director of the so-called Indian Ocean Experiment (INDOEX) with Paul Crutzen of the Max Planck Institute for Chemistry in Germany. "The local effect is well-known," Ramanathan says, "because wintertime haze can sometimes close airports in India and Pakistan for weeks. But it was not known that it had spread over the entire ocean. It stunned us to discover how pervasive these aerosols are."

In August 2002, Ramanathan and Crutzen released the first comprehensive scientific report on the phenomenon under the banner of the United Nations Environmental Programme (UNEP), "The Asian Brown Cloud: Climate and Other Environmental Impacts." Despite the dispassionate scientific language, it was a devastating document, based on INDOEX studies from a team of more than two hundred scientists in Europe, India, and the United States. Nearly 700,000 deaths worldwide are related to air pollution every year, it said, and the number could climb to eight million by 2020. The 2-mile thick cloud of pollution may already be causing the premature deaths of a half-million mothers and children under 5 in India each year, it added. The report sees a particularly ominous effect on rice production, which could decrease by 5 to 10 percent.

Dr. Ramanathan described the haze in an interview as "a complicated chemical soup" of soot, sulfates, nitrates, ash and dust, generated by emissions from coal-burning power plants, biofuel cooking and diesel engines." The source is not a mystery, but the effect on the oceans is still unknown. "The haze causes a loss of sunlight striking the surface of the sea, and we are just starting research on how that affects photosynthesis and ocean plankton," Ramanathan says. "One thing we know is that it is certainly not 'fog.' The people who call it that are in massive denial. This is not a natural problem—it was created by us."

Some of the side effects of the Asian cloud will aggravate existing problems, like the looming Third World water crisis. Aerosols block the sunlight, causing evaporation on the ocean surface, and thus interfere with natural rain patterns. "In my own work, the most worrisome effect is on the water cycle of the planet," Ramanathan says. "This is the century for water shortages, and the last thing we need is this particulate effect, but we appear to be stuck with it."

One of the reasons people have a hard time understanding global warming is its great complexity, and there is nothing simple about the Asian cloud. Barry Joe Huebert, an atmospheric chemist at the University of Hawaii in Honolulu, says that by blocking the solar radiation reaching the surface of the earth, aerosols can have a cooling effect that counteracts global warming. In the month of April, he says, the surface cooling effects of aerosols downwind of Asia are 10 percent higher than the warming caused by the release of greenhouse gases. A chart in the UNEP report shows cooling caused by the haze and heating caused by global warming roughly canceling themselves out. But that hardly makes the Asian cloud benign. Huebert told *National Geographic* that increasing temperature differences could ratchet up the intensity of storms. "One possibility is that this could cause more severe storms, more droughts and more floods."

Ramanathan agrees with Huebert that sometimes contradictory forces are at work. "The Asian plume is having a cooling effect on the surface," he says, "but at the same time it is also warming the atmosphere."

The cycle of both droughts and floods observed by Huebert is potentially catastrophic. The loss of ocean sunlight may be dramatically altering the whole hydrological cycle, a possible explanation for the severe droughts India has been experiencing. In yet another effect, aerosols are caught up in regional thunderstorms, falling back into the ocean as acid rain.

"The haze affects the optical properties of the clouds, making them brighter," says Dr. Joseph Prospero a professor of marine and atmospheric chemistry at the University of Miami and an INDOEX participant. Clouds also become more reflective and longer-lasting, a further cooling effect. Warming comes into it, too, because the pollutants are very dark and absorbing, taking in large amounts of sunlight. "When you redistribute solar energy within the atmospheric column, it has a considerable impact on the properties of clouds," Prospero says. "It's a complex system with a lot of feedbacks that are not clear yet."

Prospero adds that National Science Foundation grant reviewers had originally doubted the researchers would find anything measurable. "But the size of what we saw didn't surprise me because I've been running cruises in that area for many years," he says. "And anyone who's ever been to India knows there's a lot of pollution there. It's scary, and it's the way it will go in all of Asia."

## Pollution Travels

And the soot may not remain in Asia, because pollution does not always stay put. Scientists say that the pollution cloud could travel around the world in less than a week, carried in the upper atmosphere. Atmospheric chemist David P. Parrish of the National Oceanic and Atmospheric Administration Aeronomy Lab in Boulder, Colorado, says recent research reveals a disturbing fact: According to National Science Foundation–supported work in 2001, Asian soot particles are "piggybacking" on dust clouds blown from Asia to the west coast of the United States. "When particles run into each other they tend to stick together," he says. "It's all being transported on the same air mass."

The full extent of the problem is not yet apparent, Parrish says. "We're beginning to feel our way around the elephant," he said. "Only a relatively few episodes have been studied in detail." The net effect, Parrish adds, is to increase both global warming and coastal pollution. "The dust tends to scatter solar radiation back into space," he said, "while the soot absorbs it, heating the atmosphere."

Parrish notes that a study of springtime ozone levels in California showed that they had increased by a third between 1985 and 2002. "We can point to Asia as the likely cause, but we have no clear evidence," he said. "The increase was larger than we had expected, and it reduces the latitude we have to mess up our own air."

There is no question that dust can move across considerable distances. Dust storms are as old as time and completely natural, but our ability to track their peripatetic journeys around the world is relatively recent. New research reveals just how dramatic and far-reaching that movement can be. These are not global warming effects, but they behave in much the same way.

In 2000, a team from the U.S. Geological Survey's Center for Coastal Geology reported that several million tons of dust resulting from droughts in Africa was being transported annually to the western Atlantic Ocean, where it was reducing visibility in the Virgin Islands and causing the temporary closing of airports. "Our hypothesis is that some of the decline of the reefs in [the Caribbean] is linked to the increase in dust transport," the team said, adding that 1-inch grasshoppers from Africa had been blown to the windward islands in a 1989 dust storm. "If they can make it, think of all the other things that can make it," said Eugene A. Shinn, who led the team.

Shinn says that the biggest losses of Caribbean coral occurred in 1983 and 1987, which were also the years of the greatest dust movement. Further, in the mid-1990s, an epidemic that killed sea fans in the Caribbean was caused by a soil fungus. The fungus, Aspergillus, was believed to have come from run-off caused by deforestation, but there were also outbreaks on remote islands with no forests at all.

Two dust storms that occurred in the Gobi Desert in 1998, killing twelve people in Xinjiang, China, crossed the Pacific Ocean in 5 days, carried by the western winds that are typical for the northern mid-latitudes. They came to ground between British Columbia and California and created a small furor. Phenomena like this are described in dry scientific papers that, in rare lapses, offer some clear prose. In a study prepared by a virtual working group headed by Washington University in St. Louis, "The Asian Dust Events of April 1998," it is noted, "The most noticeable impact of the dust [which arrived in North America on April 25, 1998] was the discoloration of the sky. Human observer reports and digital photographs indicate that from April 25 onward, the normally blue sky appeared milky white throughout the non-urban West Coast."

In an effect similar to the Asian cloud, there was a marked 25 to 35 percent decrease in direct normal solar radiation measured in Oregon, and high levels of particulate matter in California, Washington, Oregon, and Nevada. In most affected states, health agencies issued air pollution advisories.

Not all the effects of high-flying dust are bad. It is not "pollution" in any real sense, and it is an ancient natural phenomenon. Although the dust particles present respiratory challenges exemplified by those public health warnings, Goddard also notes that transported dust "can be a vital nutrient source for both the oceans and terrestrial ecosystems." Iron in the dust nourishes iron-deficient oceanic regions. In a remarkable example of our holistic Earth at work, Saharan dust feeds the canopy of the Central and South American rain forest.

These effects are part of an amazing cycle. In colder periods of Earth's history, deserts expand, creating larger dust storms and more transportation. This increases iron levels in the ocean, aiding aquatic photosynthesis and absorbing more atmospheric carbon dioxide. The result is longer cool periods.

The late oceanographer John Martin hypothesized that global warming could be reversed by dumping iron in the world's seas. "Give me a half tanker of iron, and I will give you an ice age," he said. By strategically sprinkling iron in high-nutrient, low-chlorophyll zones, he said, algae blooms would be created that would take in enough carbon dioxide to reverse the greenhouse effect and cool the planet.

There are other, less-novel solutions to solving Asia's pollution crisis. Even as its forest fires raged in 2001, Indonesia hosted a four-nation summit on reducing the fires, which are often set deliberately by farmers to clear land in preparation for planting. "There are solutions," Dr. David Viner of the Climatic Research Unit at the University of East Anglia told the BBC. "Stop burning the forests, switch to less-polluting fuels, and introduce clean air technology, like scrubbers on power station chimneys. They're simple to work out. Unfortunately, they're rather more difficult to implement." Professor Ramanathan agrees. "We need fewer SUVs and more fuel-efficient cars," he says.

When politicians read hair-raising stories about global warming, their tendency is to shoot the messenger, and so it proved in India. Professor Ramanathan's studies of the Asian brown cloud were made public in the summer of 2002, and they set off their own storm cloud. Indian politicians were particularly incensed that their country was being singled out as a culprit. India's Environment and Forests minister, T. R.

Baalu, claimed that India produces particulate emissions as a "necessity," because its people are poor and must therefore burn animal dung and other readily available materials. The reaction recalled the Bush administration's dubious decision to simply excise a whole section on global warming from an Environmental Protection Agency report.

The outcry produced its own fallout, most notably a loss of further funding for UNEP's INDOEX work. The controversy is particularly distressing for Ramanathan, who grew up in India with cow dung on the cooking fire. One of the loudest protesters against the UNEP conclusions is Ramanathan's own alma mater, the Indian Institute of Science in Bangalore. Some research will undoubtedly continue, but UNEP is now using the more general phrase "atmospheric brown cloud."

PHOTO ESSAY

# Witness to a Warming World

*Gary Braasch*

Climate change is happening. I have seen it with my own eyes, and I see it right now. I have stood in the empty rookeries of displaced Adelie penguins, and felt a chill from the receding ice of the Antarctic Peninsula. I saw young black spruces growing higher than ever before on boreal hillsides in Alaska, and subtle changes transform the tundra. Near my home in the Pacific Northwest, I watch the slowly melting glaciers, and in the Andes, have rephotographed 65-year-old images of great glaciers to show them wasting away. Along the coasts, I have seen rising tides and heavy storms erode beaches. In the woods of eastern North America, I walked through spring wildflowers and spotted incoming migrant songbirds, knowing them to be arriving disconcertingly early.

I made these and other observations as part of a personal photographic project, "World View of Global Warming." I wanted to venture beyond the raw statistics, the charts and the predictions. I wanted to create an alternative to the numbers, the arguments over "who's to blame," and what palliative measures governments and corporations might be willing to take. I looked instead at the earth itself, with the eyes of a natural history photographer. Global warming and climate change have been set in motion. Ecosystems and species are already reacting. In both remote locations and familiar gardens and parks, scientists are devoting their careers to documenting the effects. This evidence, however, is largely dismissed by the Bush administration and is just beginning to be debated in Congress and by politicians. The visible effects of warming and scientific research are rarely covered by the public media.

Capturing these effects on film poses a great problem. Changes have been unfolding for 50 years or more; each year's effects are small. They are subtle and incremental, if not literally invisible. But after three years of visiting scientists at their sites and hearing their passionate concerns, working with images from the past, and documenting the meticulous record keeping of scientific fieldwork, my photographs begin to add up.

Photography's message is strengthened because global warming is revealed in the earth's most beautiful and sensitive landscapes. Treasured and threatened ecosystems and creatures are in transition. Like some early signs of heart disease or cancer in our bodies, the first effects are strongest in the extremities of our planet. The poles, the mountains, and the animals and plants on the edge of their ranges are feeling it strongly.

I have come to believe that I am documenting one of the most crucial, overarching events of the twenty-first century. As it exacerbates overpopulation and food crises, climate change may affect more people than did war in the last century. Our agriculture and use of fossil fuels are major causes of global warming. Either directly or because of our deep interconnections, everyone on earth is affected by these changes. We must try to slow them, and then adapt and live through the ones already upon us. This is an urgent story that is just beginning to be told.

One of the most striking effects of global warming is the temperature increase on the Antarctic Peninsula. The rise is especially noticeable in winter. Readings have increased 9° F over the past 50 years. Across the Peninsula, seven ice shelves like the Müller (top) and the larger Larsen are disintegrating, losing a total of more than 5,000 square miles since 1974, according to the National Snow and Ice Data Center. This warming is also affecting land glaciers. The mile-long ice cliff of Marr Ice Piedmont, Anvers Island (left), has receded about 1,600 feet since the mid-1960s. © 2003 Gary Braasch

Greenland's huge icecap, second only to Antarctica, is also showing signs of change. Outflow glaciers, which drain ice from the icecap to the sea, appear to be thinning and flowing more rapidly, as measured by NASA airborne surveys. The Jakobshavn Glacier, whose icebergs clog Disko Bay, is now 39 feet thinner and flowing more than 5 miles each year. The National Climate Data Center (NOAA) reports that 2002 saw the greatest measured surface melt of Greenland ice in 24 years of satellite records. Increasing meltwater in the North Atlantic may soon interfere with the interchange of great ocean currents such as the Gulf Stream. © 2003 Gary Braasch

Mountain glaciers everywhere in the world are receding under more than a century of rising average temperatures, and one of the most dramatic glacier withdrawals has been in the Alps. This pair of images shows a 1859 etching of the Rhone glacier in the canton of Valais, Switzerland, when the ice filled the valley right to the tiny crossroads of Gletsch. In 2001 (lower photo) the glacier was nearly out of sight, one and a half miles distant and 1,500 feet higher. In the United States, Glacier National Park is severely affected by shrinking of glaciers. All but a few of the 30 glaciers in the Northern Rocky Mountains park will be gone by mid-century. © 2003 Gary Braasch

Sea-level rise also affects the Arctic, where it frequently combines with permafrost thaw to create severe erosion. The native village of Shismaref, Alaska, a village of about 590 Inupiats perched on a sandy barrier island on the northwest shore of Seward Peninsula, has failed to halt the rising Bering Sea. Shore erosion of the narrow spit has been severe since the 1950s, and protective barriers have been ineffective. Townspeople voted in 2002 to move their village to higher, more-protected ground away from the ocean, but also far from their traditional fishing and sealing sites. © 2003 Gary Braasch

Around the world, sea level is rising, due to warming of the ocean and the inflow of melting glaciers. The average rise in the twentieth century was 6 inches, and the rate will have increased to 1 to 2 feet by 2100, threatening every coast with higher waves and stronger surges during storms. Even on a clear day, this sandy shore at Vero Beach, Florida, is being washed away by a normal high tide. This casts doubt on the safety of populations that crowd the ocean coastlines in Florida and globally. Rising sea level is also driving sea water into the Everglades, inundating mangroves, flooding estuaries, and submerging all low-lying islands. © 2003 Gary Braasch

Geophysicist Tom Osterkamp indicates ground-level subsidence in the 15 years since he installed this temperature probe pipe near Denali Park. Alaska permafrost temperature has increased by up to 2.7° F since 1980, when Osterkamp began measuring ground temperature at 30 sites. Uneven thawing of permafrost is causing widespread ground subsidence. Serious effects include forest damage, sinking roads and buildings, eroding tundra riverbanks, changes in tundra vegetation, and increased carbon dioxide and methane emissions from thawed tundra. © 2003 Gary Braasch

In Alaska the very ground is changing as the climate shifts to the highest temperatures in more than 400 years. Scientists have been intensely studying the tundra since the 1970s, noting an earlier meltoff of winter snows, change from absorbing carbon dioxide to giving off some CO2 and methane, and increased growth in shrubs. The change in tundra seasons and plants may affect Alaska's famous caribou herds, whose birthing evolved to coincide with spring plant growth. Arctic climate warming also is changing patterns of snowfall, making winter forage harder to find for the caribou. © 2003 Gary Braasch

Ornithologist Bill Fraser stands on the smooth pebbly surface of a former large colony of Adelie penguins in the Antarctic Peninsula. This nesting site has been used by fewer and fewer penguins over recent years. Analyzing climate data, island topography, and breeding statistics, Fraser believes regional warming caused the loss of half of the 16,000 nesting Adelies on this island. Part of the cause is reduction in winter pack ice, where penguins' choice prey, tiny, shrimplike krill, develop. © 2003 Gary Braasch

David Walker of the Dungeness Bird Observatory, Great Britain, holds a "chiffchaff," a warbler whose yearly migration and nesting are much earlier now than in the mid-twentieth century. England has been a center of nature observation by scientists and amateur naturalists for hundreds of years. These records show that more than 100 birds now migrate and nest earlier and/or have moved north to take advantage of warming climates. Similar data throughout Europe and in North America confirm that birds are reacting to global warming. © 2003 Gary Braasch

Butterflies are sensitive monitors of climate change because of their dependence on plants, and because of their mobility. Comparing old records of the Edith's checkerspot butterfly against current habitat from Mexico to Canada, biologist Camille Parmesan found that the insect had moved its habitat northward in response to climate warming. In Europe, Parmesan found 66 percent of butterflies have moved north during the twentieth century. Plants are also leafing and blooming a week or more earlier throughout much of the Northern Hemisphere, and predictions are that many plant/animal associations will be upset. © 2003 Gary Braasch

Dr. George Divoky has documented the climate-mediated rise and decline of a colony of black guillemots on a barrier island in the Arctic Ocean. Successful nesting was made possible by the slow retreat of Arctic sea ice from around the island, according to Divoky's observations, and more than 225 pairs nested on Cooper Island in the late 1990s. But the ice has continued to move away from the island and with it the prime prey of the guillemots, Arctic cod. The seabird colony has declined to fewer than 150 nests. © 2003 Gary Braasch

A large number of polar bears have been foraging on land in recent years, a probable result of the retreat of Arctic Ocean near-shore ice. In studies in Canada and Europe, the bears are shown to suffer from nutrition and denning problems when the ice withdraws from shore until very late in the fall. Changes to birds and bears are just a part of the effects from the 14 percent reduction in permanent Arctic sea ice since 1970. The trend is that the Arctic Ocean could become icepack-free during the summer by 2100. © 2003 Gary Braasch

Summer sea temperature has risen more than one degree F since 1930 in a tide pool at Hopkins Marine Laboratory in Monterey, California. The resurvey showed many warm water tidepool animals have increased, while those favoring colder water have decreased. Rising ocean temperatures and intensified El Niño events are also devastating to coral reefs, causing the coral to lose their symbionts—bleaching—and encouraging suffocating growth of algae. © 2003 Gary Braasch

Even in the tropics, a warming atmosphere is desiccating some of the habitat and causing changes in flora and fauna. The most celebrated scientific evidence is the disappearance of the golden toads, *Bufo periglenes*, at Monteverde Cloud Forest in Costa Rica. Each year Dr. Alan Pounds and others search for the distinctive orange amphibian in its restricted habitat along a narrow, fog-bound ridge. About 1,500 toads were sighted in 1987. But now the breeding pools remain empty — the toad has not been seen since 1991 and is feared extinct. © 2003 Gary Braasch

Chicago in August 1995 broils under a thick blanket of smog in a record heat disaster that killed more than 700 people. Climate predictions are that punishing heat waves will become more common. This was borne out in August 2003, when more than 55,000 people died in record hot spells in Europe and India. Tropical diseases and asthma are spreading.

A warming atmosphere is less stable, creating stronger storms, longer droughts, and heavier downpours. Insurance companies paid out a record $60 billion on weather disasters in 2002 and $55 billion the year before, according to the reinsurer Munich Re. © 2003 Gary Braasch

# Part Two

# Ecosystems in Trouble

> It is an almost foregone conclusion if we do nothing. As a result of the rapidly rising global temperature, corals will largely become remnant organisms on reefs that once had extensive stands of them.
>
> —Dr. Ove Hoegh-Guldberg, Director, Center for Marine Studies, University of Queensland

## Chapter Six

# Alaska and the Western Arctic: The Ice Retreats

*Kieran Mulvaney*

We squint through the fog, trying to make out the shape we know we should be able to see by now. The charts say it is there, and the radar shows it clearly, but to our eyes there still is no sign of the island that lies just 2 miles ahead. Then, slowly, the mist parts, and our blindness gives way to incredulity. It seems almost impossibly barren, a boulder-strewn piece of rock in the middle of freezing, gray ocean, with a ram-shackle collection of huts and houses standing defiantly at its base. Gazing at the unfolding scene from the protection of the ship's wheel-house, we look at each other in disbelief. People really live here?

Welcome to Little Diomede Island, a barren island outpost of the Last Frontier, a few miles off the coast of Alaska in the cold and hostile Bering Strait. The island is home to about two hundred Inupiat Eskimos, descendants of settlers who spread east from the nearby Siberian coast perhaps 2,000 or 3,000 years ago. If the island itself seems uninviting, the attractiveness of the area to both those early settlers and today's inhabitants is easy to see: The waters of the region support a relative bounty of marine life, and the Bering Strait acts as a bottleneck through which migrating marine mammals must pass on their journey from their warm-water breeding areas to their summer feeding grounds in the Arctic. For a culture focused on hunting whales, walruses, and seals, that made—and makes—Little Diomede desirable real estate.

Life on Little Diomede—and for Inupiat and Yupik Eskimos elsewhere in the Bering Strait and beyond—was frequently challenging and harsh, with periods of bounty countered by the brutality of the environment. Perhaps predictably, however, that way of life encountered its biggest challenge with the arrival of Americans and Europeans, beginning in earnest in the nineteenth century. American whalers and sealers devastated the populations of whales and walruses on which the Inupiat depended, and introduced diseases and alcohol. Missionaries demonized traditional spiritual and religious beliefs and practices; speaking Inupiaq instead of English was sometimes prohibited.

The Cold War inflicted particular injustice on the people of Little Diomede: Just a mile or two east of the International Date Line, Little Diomede is in American territory, part of the state of Alaska; neighboring Big Diomede is on the other side of the date line and under Russian—formerly Soviet—control. Before such boundaries, interactions between the two islands were close and constant, with both isles boasting members of the same families; then the Iron Curtain bisected the Bering Strait, and the people of Big Diomede were forcibly assimilated into mainland Soviet society.

Today, Little Diomede faces a new threat, one pithily summed up by resident Anthony Soolook, Jr., as he and I stand on deck and look across the water at his home in the near-distance, and at countless boulders that loom ominously above the little village. Anthony points to places where the boulders have become dislodged, carving paths through the shallow soil. "Last night," he says, "I had a dream, that the boulders came down and smashed our village."

There have always been landslides on Little Diomede. The soil is thin and loose, providing little support for the rocks on the surface. However, in the past, it has all been anchored in place by permafrost, the layer of permanently frozen soil that is a distinguishing feature of polar and alpine environments. Now, however, the temperature in the area is rising, the permafrost is melting, the ground above it is sinking, the rocks are becoming dislodged, and landslides are becoming more frequent and severe. Shortly before our visit, one such slide destroyed a carefully constructed pathway up the mountainside. "In the past, when that happened, we would rebuild the path," says one village elder. "Now, I don't think we will bother."

It is not just landslides, and it is not just Little Diomede. Along the coasts of the Bering, Beaufort, and Chukchi Seas, residents of Alaska Native villages tell similar stories: of warming temperatures, changes in snowfall, reduction in the thickness and extent of sea ice, and potentially negative consequences for a traditional subsistence-based way of life:

"There used to be heavy snowfall here; there used to be three feet of slush where we walked and now I don't see it any more."

"We hardly got any snow until November. Usually we have our first snowfall around the end of September. During the summer months we have clouds and rain and drizzle. Now there's hardly any clouds or rain and drizzle, there's more sunshine. It's a lot warmer than before."

"The most change I've seen is how thin the [sea] ice is getting. Year by year."

"The ice used to be five to six feet thick. The last couple of years it's been four, four and a half feet. That's a foot, foot and a half, and that's a pretty substantial difference. . . . Break-up seems to come quicker. Sometimes a couple of weeks, sometimes as much as a month sooner. Freeze-up was as much as a month late."

"The thing that I notice when I walk out on the tundra is that it isn't as spongy as it used to be. Now I can hear it crackle when I walk on it, and it's dry. It's real dry. Whereas before some places I did not go because they're too wet, now [in] the areas where lakes used to be, all the plants are dried."

To Melanie Duchin, Alaska representative of the environmental organization Greenpeace, which recorded these testimonies and published them in a 1998 report, the comments of the Yupik and Inupiat demonstrate that "climate change is not just a theory, it is a reality. It is happening now, and it is having a very tangible effect on people today."

Such bravura statements might elicit derisive snorts—or, at the very least, raised eyebrows—from skeptics. It is not only that Greenpeace would be expected to make that kind of claim, but also because many researchers have traditionally dismissed the reliability of anecdotal observations such as these. But almost without exception, the evidence presented to Greenpeace and others by Alaska Natives correlates perfectly with the observations and predictions of scientific researchers.

For one thing, the Arctic region is indeed, as Native testimonies claim, warming: compared to the rest of the world, significantly and rapidly. Global temperatures have increased on average by 1°F over the past 100 years, but in the Arctic, according to the Intergovernmental Panel on Climate Change (IPCC), "The magnitude of warming is about 5°C [9°F] per century." Growth rings in Siberian larch from western Siberia show that the twentieth century was the warmest in more than 1,000 years, and that the region has experienced a steady temperature increase for 150 years. A 2002 paper in the journal *Science* concluded that the average Arctic surface air temperature in the twentieth century "was exceptionally high compared with the previous 300 years."

On the basis of such measurements, the IPCC predicts a range of climate effects in the Arctic, from ecosystem changes to diminishing sea ice. Such images are largely couched by the IPCC in hypothetical terms, predicting what models suggest may yet come to pass. Those who deny global warming's reality question such assumptions, citing alternative models and more optimistic interpretations. But as the people of Little Diomede are all too well aware, the changes brought about by rising temperatures in the western Arctic are far from theoretical. And some of the most striking images of the very real changes that are taking place in the region can be found within the boundaries of Alaska's largest city.

## Dying Forests

The Seward Highway is the main thoroughfare south from Anchorage. It runs down the Kenai Peninsula to the small, attractive fishing town of Seward, diverging en route into the Sterling Highway, which leads to the equally small, and perhaps even more stunning, fishing town of Homer. It is a devastatingly beautiful stretch of highway, alternately revealing stunning mountain vistas, the still waters of Cook Inlet, and long stretches of coniferous forest. A closer look at the trees that cover the hillsides, however, reveals that all is not well with that forest. Huge stretches of them are dry and colorless, weak, dying or dead. They are victims of the spruce bark beetle, which has killed Sitka-white spruce hybrids over an area in excess of 3 million acres of south-central Alaska. The beetle uses sensitive antennae to detect subtle chemical signals from spruce that are slightly weakened, stressed, or diseased; these trees, less able to resist the beetle's attack, are its favored targets. After laying its eggs in its victim, the beetle then itself emits hormones, attracting additional beetles to the site. Soon, the tree is overwhelmed by the attentions of literally thousands of beetles, and is killed.

On drives along the Seward and Sterling Highways, and on hikes through Chugach State Park that overlooks the Anchorage Basin, I have seen this widespread devastation repeatedly. (Indeed, the spruce tree that once stood proudly outside my cabin ultimately succumbed to a siege of bark beetles, whereupon it was cut down, ultimately providing valuable warmth from my fireplace over the winter.) Now, one bright summer afternoon, on an island in Prince William Sound—the beautiful stretch of water most famous, tragically, for the obscenity visited upon it in 1989 by the oil tanker *Exxon Valdez*—I come face to face with the bark beetle's unwitting accomplice in the assault on Alaska's conifers. I am with Dr. Glenn Juday, a forest expert with the University of Alaska, Fairbanks, and we are in search of the western black-headed budworm. The search does not take long; with a triumphant "Aha!" Juday beckons me to a nearby spruce, and points to the tell-tale sign—spruce needles bound tightly shut—of the western black-headed budworm. After hatching, budworm larvae burrow their way into the buds of Sitka spruce and suture them with their silk. After feeding on the buds through spring, the larvae hatch as caterpillars in the summer and feast on the trees themselves.

All of this is, on one level, entirely natural. Both the spruce bark beetle and the western black-headed budworm are natives of south-central Alaska. But warmer temperatures have made conditions increasingly favorable for them both: Warm summers have enabled the bark beetle, for example, to halve the length of its life cycle from two years to one, effectively doubling its population size. At the same time, warmer, drier conditions are weakening the trees and making them more vulnerable to insect attack. Such is the scale of the devastation that, according to Juday, the "entire forest system is dying, and the question is what will replace it." According to some researchers, it is a scene that is likely to be played out across the Arctic; some say as much as half of the world's boreal forest could be gone within decades.

## Starving Bears

A little more than a month after looking for black-headed budworm with Juday, cruising up the west coast of Alaska, and recording the testimonies of the residents of villages such as Little Diomede, I am standing on the deck of the icebreaker *Arctic Sunrise*, anchored just off Barrow on Alaska's north coast, and, with most of the ship's crew, watching a black dot in the distance swim slowly toward us. At first, we assume it is a seal; on closer inspection, however, it reveals itself to be a polar bear. Eventually, it arrives off our ship, head in the air, sniffing longingly at the scent of freshly baked bread wafting from the galley. It clambers on to an ice floe in an attempt to get closer, and that is when we notice: It is alarmingly thin.

It is unusual enough to find a polar bear swimming through so many miles of open ocean: Polar bears are creatures of the sea ice, across which they wander for miles in search of their walrus, seal, and sea lion prey. Eventually, the bear gives up and swims slowly and sadly away, but it does not take long to discover exactly why it was laboring through open water instead of patrolling the sea ice. Shortly after entertaining our visitor, we set off in search of the ice edge, but despite our traveling far north of where it should be, it takes days before we find it.

Then we cruise west, far north of Alaska, and all the way to Wrangel Island, off the Arctic coast of Russia.

As the wind whips off the ice floes, I hunker down in my cold-weather gear, sheltering in the leeward side of the bridge wing alongside Drs. George Divoky and Brendan Kelly, of the University of Alaska in Fairbanks. Each is on board the icebreaker to lead separate teams of researchers: Kelly's team conducting a survey of walruses, Divoky and his assistants counting seabirds. For Kelly, it provides an opportunity to conduct the first such survey in this region for several years: an especially important task, given the concerns about retreating sea ice and its possible implication for walruses, which "haul out" and rest on the sea ice.

Divoky hopes his survey will provide potential supporting evidence for a thesis he has nurtured over three decades of research on Cooper Island, just north of Barrow. Year after year, he has set up camp here, and at times it has been an uncomfortable, discomfiting undertaking. Encounters with polar bears are frequent, and Divoky recalls one particular experience when he heard something in the middle of the night and woke up the next morning to find bear tracks leading to the edge of his tent, just inches from his head. But Divoky has persevered, and as a result, he says, he has gathered evidence of what is "among the first documented biological effects of climate change in the Arctic."

The focus of Divoky's research is the black guillemot, a species of seabird. Found throughout the higher latitudes of the Northern Hemisphere, black guillemots have not been common in northern Alaska since the seventeenth century, when global temperatures began to fall and snow cover increased. That is relevant because, says Divoky, the guillemots require 80 days between nesting on the ground and hatching their eggs, but over the last 300 to 400 years, there have rarely been 80 consecutive days on which the ground in northern Alaska was snow-free. Until now.

In 1972, Divoky discovered ten pairs of guillemots on the island, nesting in the remnants of ordinance that had been blown up in the 1950s. Over the next 18 years, the number of birds soared, peaking at 225 pairs in 1990. The increase was fueled, says Divoky, by warmer temperatures causing earlier spring snowmelt, allowing for a greater number of days in which the ground was snow-free and providing more opportunities for the birds to nest.

After 1990, however, guillemot numbers began to decline again; the colony is now below one hundred pairs once more and continuing to fall. Those birds that remain are lighter and thinner; mortality rates are higher; and the number of immigrant birds from Wrangel Island, which Divoky believes is the source colony, is dwindling. The cause once again, Divoky believes, is climate change.

This time, however, the problem is that the same warm temperatures that have melted snow on the ground are also melting sea ice, which is as important for seabirds such as the black guillemot (which use it as a place to rest during long flights, and which feed on the Arctic cod which shelter beneath the floes) as for polar bears, seals, sea lions, and walruses.

## Nice Weather for Mosquitoes

East of Cooper Island, in the Canadian province of Nunavut, a different species of guillemot is also showing the effects of climate change, in a most unexpected way. Since 1990, Anthony Gaston of the Canadian Wildlife Service and colleagues have been studying a breeding colony of Brünnich's guillemots on Nunavut's Coats Island. The guillemots have the particular misfortune to be located near an area of high mosquito abundance, and from 1997 to 1999, note the researchers, mosquito numbers were so high that more than fifty mosquitoes at a time would swarm over each foot of the unfortunate birds. Eventually, the harassment and discomfort became so great that many of the birds were forced to abandon their breeding sites, whereupon many of the abandoned eggs were eaten by gulls As a result, Gaston and colleagues recorded in a 2002 paper in the journal *Ibis*, the "proportion of eggs lost was about twice as high, on average, on days when mosquito abundance was high." Additionally, evidence from necropsies conducted on a number of dead adult birds "was consistent with mortality in response to a combination of temperature and mosquito attacks, consequent loss of blood and water (through panting to increase evaporative cooling) causing dehydration and heat stroke." They conclude that "significant global climate change may contribute to increased mortality and reproductive failure in Brünnich's guillemots through increased mosquito activity. Insects pass through larval stages more quickly at higher temperatures, becoming adults earlier; this appears to be what we are seeing with the mosquitoes at Coats Island."

The experience of the Brünnich's guillemots of Coats Island shows how just one or two excessively warm years can be enough to prompt slight changes that can prove devastating to small populations. Elsewhere in the Arctic reaches of Nunavut, there is a similar story.

In 1996, biologist Frank Miller landed at his base camp on the shore of remote Bathurst Island, far north of the treeline, in the bleak regions of Canada's High Arctic. For 8 years, he had been returning to this place each summer to study the Peary caribou—a smaller cousin of the familiar caribou and reindeer that inhabit much of the Arctic.

This year, however, proved different from others. Whereas before he had been able to estimate the number of caribou in the region by counting the animals he saw during helicopter flights over the area's ten islands, this time he hardly saw any live caribou at all. Instead, seemingly everywhere, there were carcasses: hundreds of them, of musk oxen as well as Peary caribou—most lying on the ground, but some still standing, frozen into the sea ice. By the time he had finished his count, he had been able to find only ninety-one Peary caribou and ninety-seven musk oxen, from which he extrapolated that, in an area covering 11,000 square miles, there were a mere 500 caribou and about 430 musk oxen. This represented declines of 85 and 70 percent, respectively, from the previous year's figures.

Two years later, he and fellow biologist Anne Gunn provided an even more sobering picture, when they could find only forty-three caribou and seventeen musk oxen. What had happened, the biologists surmised from the evidence before them, was that unusually warm temperatures in the western Arctic had brought warm, moist air masses from the south, resulting in the normally dry High Arctic receiving unprecedented levels of snow, followed by unheard-of freezing rain. Layers of alternately freezing and melting snow, and the freezing rain, blanketed the ground where the Peary caribou sought to forage, cutting them off from their food source and causing them to die of starvation. As Gunn explains it to me: "The caribou expended so much energy trying to dig through the snow and ice to get at the food below that they just ran out of energy and then they died."

## Seabirds in Peril

In 1997, conditions in the Bering Sea were unusually warm, windless, and cloud-free. Sea-surface temperatures were the highest ever recorded. And the ocean's top layer did not mix well with lower levels, leading to a rapid depletion of nutrients in the upper layers and the largest bloom of coccolithophore algae ever seen in the Bering Sea. That same summer, as many as 200,000 short-tailed shearwaters—approximately 10 percent of the regional population of this seabird species—died, apparently of starvation. The birds that survived were lighter than they had been a year earlier. The events were not coincidental: With light levels lower beneath the bloom, the seabirds may have had trouble seeing their fish prey. They would likely have had difficulty reaching them as well, as the warmer temperatures seemingly prompted the fish to seek refuge in deeper, cooler waters where the shearwaters could not reach them. The following summer, conditions had somewhat abated, but the bloom remained, and was large enough to be photographed by NASA satellites. And there was another mass mortality of seabirds, this time common murres, tens of thousands of which died from Cook Inlet west to the Aleutian Islands.

The algal bloom and associated seabird deaths were but the latest in a series of unusual events and wildlife declines to have struck the Bering Sea and Gulf of Alaska in recent decades. Several types of fish-eating birds have undergone declines of as much as 50 percent in Prince William Sound, although populations of bottom-feeding birdlife in the same region have remained stable or increased during that time.

Populations of several species of waterfowl, among them Steller's eider, spectacled eider, and various species of loon, have undergone precipitous declines, sometimes in excess of 50 percent, on their breeding grounds in the Yukon-Kuskokwim Delta; seabirds such as common murres, thick-billed murres, and red-legged and black-legged kittiwakes have also declined significantly in the western Gulf of Alaska and Aleutian Islands.

Steller sea lion populations in the Bering Sea and western Gulf of Alaska have declined by approximately 80 percent. In parts of the Gulf of Alaska, harbor seal numbers have dropped by as much as 90 percent. Sea otter numbers in the Aleutians have fallen by 70 percent since 1992 and, in some areas of the archipelago, at least 95 percent since the 1980s.

According to a 1996 report by the National Research Council (NRC) of the National Academy of Sciences, the region is responding to a series of natural and anthropogenic impacts that date back decades and even centuries. According to this "cascade hypothesis," large reductions in whales and some fish as a result of over-exploitation increased the amount of food available for other fish and invertebrates, so that by the 1960s and early 1970s, the Bering Sea ecosystem changed from one dominated by capelin to one dominated by pollock. This change was intensified in 1977 by what is known as a "regime shift," in which the region's sea temperatures increased, to the benefit of groundfish (except for cold-water species such as turbot) and the detriment of species such as capelin and sand lance.

According to the NRC thesis, as a result of this regime shift, Steller sea lions, which previously fed on capelin and herring, were deprived of their primary food source and forced to subsist on the less nutritionally valuable pollock, initiating the demise of that species. Environmentalists and many researchers argue that the sea lions' decline is exacerbated, or their recovery impeded, by intense fishing for pollock.

A 1998 paper in *Science* argued that the decline of sea lion populations was forcing orcas, which normally fed on the pinnipeds, to switch their attentions to sea otters, and that it was this increased predation that was largely responsible for the sea otters' disappearance in the region. A further result is an increase in the sea urchins on which sea otters prey, and consequent deforestation by the urchins of kelp beds in the region. This plethora of natural and human-induced impacts makes it difficult in many instances to determine the precise impacts of climate change from other human activities, or from long-term natural climatic fluxes. Opponents of measures to counteract global warming seize on those difficulties as arguments for inaction, but in fact the opposite is true. It can take relatively little for a population or ecosystem to feel the impacts of warming temperatures—as the examples of the Brünnich's guillemots on Coats Islands, Peary caribou on Bathurst Island, and short-tailed shearwaters in the Bering Sea demonstrate clearly. In an ecosystem that is already stressed or in flux, the effects of any such change are only likely to be magnified, with unanticipated consequences. And in the meantime, while the debate about global warming continues, there are, as we have seen, several instances in Alaska and the Arctic where its reality is becoming only too clear—and that is something that Alaska Natives, in particular, know only too well.

## Ominous Increases

Global warming in Alaska is not uniform; indeed, portions of the eastern Arctic are undergoing a slight cooling. But in the western Arctic, and particularly in Alaska and the Bering Sea region, the changes are pronounced.

According to the Alaska Climate Research Center, annual average temperatures in Alaska increased by 2.69°F between 1971 and 2000; spring temperatures across the state during that period increased by 4.23°F and in Barrow on the northern Arctic coast by 6.97°. In Fairbanks, Alaska, the fall and winter of 2002 to 2003 were the warmest ever recorded, with temperatures 3° above normal in September, 8° above normal in October, 14° above normal in November, and 11° above normal in December; for the year, temperatures were 2.9° above the 1971 to 2000 average. The start of the 2003 Iditarod sled dog race was moved several hundred miles north of its usual site outside Anchorage because of a lack of snow.

A 1998 report by the Center for Global Change and Arctic System Research at the University of Alaska Fairbanks argued that in the past decade, climate change in Alaska has resulted in, among other things, declines in some salmon stocks, an increase in human health problems as new diseases move north, and increased forest fires. According to the IPCC, Northern Hemisphere annual snow cover extent (SCE) has decreased by about 10 percent since 1966; longer regional timescales suggest that Northern Hemisphere SCE over the past decade has been the lowest in 100 years.

The IPCC notes that other changes in terrestrial ecosystems include "northward movement of the treeline, reduced nutritional value of browsing for caribou and moose, decreased water availability, and increased forest fire tendencies. There are altered plant species composition, especially forbs and lichen, on the tundra." And, as the people of Little Diomede can attest, permafrost is warming and starting to melt. The IPCC notes that "multi-decadal increases in permafrost temperature have been reported from many locations in the Arctic, including northern and central Alaska, and Siberia"—although it again offers the caveat that such increases are not uniform and that parts of the eastern Arctic have experienced a permafrost cooling.

As with the land, so it is with the sea. The IPCC notes, for example, that: "Data gathered from submarines indicate that sea-surface temperature (SST) in the Arctic basin increased by 1 degree C [1.8°F] over the past 20 years, and the area of warm Atlantic water in the polar basin increased by almost 500,000 kilometers [193,000 square miles]. Field measurements in 1994 and 1995 showed consistent Arctic seawater warming of 0.5 to 1 degree C [0.9 to 1.8°F], with a maximum detected in the Kara Sea." (The IPCC does note, however, that it "is not yet clear whether these changes are part of low-frequency natural variability or whether they represent the early impacts of long-term climate change.")

A series of studies have pointed to decreases in the extent and thickness of sea ice—literally, the frozen surface of the sea—in the Arctic. Throughout the Arctic, sea ice extent diminished by about 3°F per decade between 1978 and 1996: That is about 143,000 square miles—an area almost as large as Montana—every 10 years. In the Atlantic part of the Arctic Ocean, the extent of summer sea ice has shrunk by 20 percent over the last 30 years, and spring sea ice extent in April in Nordic seas has been reduced by 33 percent over the last 135 years. A 1999 study using sonar data from nuclear submarines argued that the sea ice that remained in the Arctic was as much as 40 percent thinner than it was two to three decades ago.

The IPCC offers a number of predictions for future temperature increases, the resultant changes in Arctic environments, and the consequences for the region's biota. It suggests that, ultimately, Arctic land regions will on balance receive considerably more snowfall in winter (although that snowfall will melt more quickly) and that the climate will be markedly warmer, with warming most pronounced over North American and Eurasia. Spring is likely to arrive, on average, 7 days earlier, and fall a week later. The tundra may decrease by up to two-thirds of its present size. The Arctic Ocean, too, will likely become warmer and wetter. Sea ice will continue to thin and retreat, to the extent that by 2050, sea-ice cover in the Arctic Ocean may be reduced to about 80 percent of the area it covered in the mid-twentieth century. Such sea-ice declines may well have negative repercussions for many species of marine wildlife, including seabirds and marine mammals: seal species, for example, that use the sea ice as a platform on which to rest, and polar bears that prowl the ice to prey on seals.

Changes as a result of rising temperatures are much more than an inconvenience to the peoples of the Bering, Beaufort, and Chukchi Seas. They are a profound threat to the very way of life they have practiced for centuries. Thinning and retreating sea ice, for example, makes it dangerous to hunt walruses, seals, and whales; in 1998, whalers from the village of Wainwright had to be rescued after the ice floe they were on broke up and drifted out to sea. Sea ice also provides protection for coastal villages from the storms that ravage the Arctic coast—and which themselves are predicted to increase as a result of climate change. Without the protection the sea ice offers, those beaches are subject to erosion, and the villages are at risk of elimination. Gambell, a village on a gravel spit on St. Lawrence Island in the Bering Strait, is being repeatedly moved inshore as a result of erosion; residents of the nearby village of Shishmaref attempted to salvage their home after a series of powerful storms in 1997 before eventually electing to abandon it altogether.

Alaska Natives see these signs, and they note the inaction of politicians so far away in Washington, D.C., and they wonder if, yet again, they will be abandoned to their fate, forced to endure the consequences of actions not of their own making.

Back on Little Diomede, Anthony Soolook, Jr., sums it up. "Our island is falling apart," he says. "But who cares? Who will come and help us, all the way out here?"

Chapter Seven

# The California Coast: Marine Migrations and the Collapsing Food Chain

*Orna Izakson*

California's famous coast, the golden shore, the stuff of songs, stretches roughly 1,100 miles from the dry Mediterranean climes of the south to the lush rain forests of the north. In sunny San Diego, weather forecasts are notoriously boring; according to legend, they are sometimes recorded days or more in advance. In the north, redwood and Douglas fir forests drink in the driving rain and tickle water out of the fog, seeing little sun for months out of the year.

In between, plant and animal communities hover in their preferred ranges in overlapping steps, slowly shifting in the same way habitats shift as you climb a mountain, or as you look higher along a tidal edge. Because of this broad spectrum of varied ecological zones, California boasts some of the greatest diversity of all the fifty United States. Its sheer length also makes it a perfect place to patiently, slowly, and meticulously observe the way nature responds to a warming world.

An hour past dawn, the central California sun burns through the morning haze. The waters of Monterey Bay are rising after a 6:21 A.M. low tide. Rafe Sagarin, curly hair tousled by the unruly wind, gingerly crosses the exposed granite tidepools like a teenager negotiating a cluttered bedroom floor. This intertidal tangle is a library to Sagarin, who knows precisely where reefs of tube snails set their mucous nets, striped sunburst anemones open their tentacles to the tides, and barnaclelike limpets farm algae for their supper.

Sagarin can also tell you that this tidal community is not what you would have seen in the 1930s. In fact, it looks a lot more like southern California. Sagarin knows because he has counted every critter along a line between two brass bolts in the rocks, and compared them to a similar count done 60 years earlier. That change is so big, so obvious, and so important that an article he and colleague Sarah Gilman wrote about it was published in the prestigious scientific journal *Science* while they were still undergraduates. That research, which used their counts combined with temperature and other measurements going back three-quarters of a century, found that the most likely cause for the massive changes in the kinds of species there was warming temperature. It is exactly what one would expect to happen as the world's climate changes.

In 1993, Sagarin and Gilman were juniors at Stanford University, among two-dozen students who took a term away from the main campus to study marine biology at the university's Hopkins Marine Station on the Monterey Peninsula. Upon arrival, the undergraduates were treated to presentations about areas of research at the station, and projects they would have the opportunity to participate in.

It was not what Sagarin expected.

"I came down here and I had a naïve view of marine science," he says. At twenty, he had thought the discipline consisted of learning about the animals that inhabit the different niches along the salt-water edges. But at Hopkins he found that most people were molecular biologists or neurobiologists, considering the marine species on a tiny, chemical scale. Sagarin says he did not understand what most of the faculty were talking about when they discussed the minutiae of their work.

One of the faculty members that spring was Chuck Baxter, a marine ecologist who had worked from the Hopkins station since the 1970s. Deep-voiced and white-bearded, Baxter looks like a silver-haired William Shakespeare with a dash of Ernest Hemingway thrown into the mix. The professor's seniority and drive made him a central figure in such projects as helping found the now-famous Monterey Bay Aquarium and developing a major film documentary to bring the complex lives of snails and clams to a public that sees them mainly on dinner plates.

Pulling no punches, Baxter offered the students the same challenge he had offered their predecessors for 5 years. He wanted to know "why there were so many damned *serpulorbus*"—a prevalent tube snail from southern California—in the Monterey area, Sagarin recalls. The tube snail covered everything down south, but until a few years earlier had been nearly impossible to find from Hopkins north.

Baxter knew the tube snails from the 1950s, when he was an undergraduate at UCLA and later as a graduate student living in Santa Monica. *Serpulorbus* covered every surface, from rocks to docks, that were splashed by Pacific waves.

Like most such animals, larval *serpulorbus* float through the water, eventually landing on and cementing themselves to rocks or other surfaces that are intermittently inundated by the tides. Once settled into place, they begin growing long, tubular shells, looking rather like the extended blur you see when something runs in slow motion. As the *serpulorbus* population grows, the overlapping tubes completely cover almost any surface that will stand still long enough, like calcified kudzu. The reefs they form help other species move into areas where they would otherwise have no toe hold. Sea urchins, for example, cannot latch onto the tough granite of the Hopkins intertidal area. But they can wedge themselves into *serpulorbus* reef and become established that way.

Researchers in the 1960s who came to Hopkins looking for tube snails to study were disappointed, and had to head south for their work. When Baxter arrived in 1974, he saw a few *serpulorbus* in the rocky intertidal area outside the station's offices. "I guess you could find one every 10 feet," he recalls. When his research took him up to Half Moon Bay and Pescadero, about 40 miles to the north, he did not see any *serpulorbus* at all.

That started to change within a few years of Baxter's arrival.

"Sometime in the early 1980s, the numbers began to pick up," Baxter says. "By the mid to late 1980s they were very abundant." The tube snails completely covered the rocks behind the Monterey Marina's breakwater, just as they had in southern California 40 years before.

The new abundance of *serpulorbus* was not the only change Baxter had noticed in his decades at the station. Earlier researchers had documented the seaweed cover on the rocks; by the late 1980s much of that was completely gone or had shifted substantially, taking over what had been habitat for barnacles and other critters. Baxter also saw that one predatory snail in the whelk family, *ocinebra*, seemed to have increased; it was another species he knew from his southern California days.

While casual observations make interesting stories, they are not the same as rigorous science. They are, Baxter says, "anecdotes that you can't do much with." To turn his eyeballed observations into respectable research would take a concerted effort to actually measure any changes, compare them with other kinds of data about the physical environment, and then try to determine the factors that might have pushed the change forward.

Baxter had a couple of starting points. He had found a small handful of older studies documenting all the plants and animals along specific sections of the rocky tide pools just downhill from the Hopkins station's lawn. The station had been collecting data including air and water temperature for decades; some statistics even went back as far as 1917. And the intertidal area between Hopkins and the open waters of Monterey Bay was California's first marine reserve, protected from fishing and other extraction since the 1930s, so the species found there would be essentially unaffected by human tastes.

"For about five years it had become very obvious to me that some very dramatic changes had taken place and were continuing to take place in the intertidal off Hopkins," Baxter says. That was the source of his challenge to the students who came for the undergraduate research course each year. What had changed? How much had it changed? What factors were associated with those changes? And, did the changes in snails and the seaweed simply represent a shift in a few individual species, or was something broader and more significant happening?

For years, the students were not interested. Like the other researchers at Hopkins, they favored quick, experimental science—the molecular or neurobiology—over tediously counting critters in a tidepool. Most looked at Baxter's suggestion as research unlikely to yield spectacular findings.

It was not until Baxter's last semester as a full-time member of the Hopkins faculty that he got a bite: Rafe Sagarin and Sarah Gilman.

They started with a study done in the 1930s by a Stanford Ph.D. candidate named Willis Hewatt, who had pounded heavy-duty brass bolts into the slow-eroding granite and meticulously counted every anemone, whelk, snail, sea star, barnacle, limpet, and other critter along the 108 yards in between.

The area where Hewatt did his work is a little triangle of pools made up of huge granite rock and ending at a bouldered point, streaked with white from the seabirds it hosts. The outer rocks provide some shelter from the more turbulent—and cold—waters of Monterey Bay, about 100 yards from shore. It is a perfect spot for looking at the creatures that thrive in the intertidal zone, the band along the rocks that is submerged at high tide, exposed at low tide, and wet by spray and waves in between. In the same way zones of life change as you climb a mountain—or as you drive north or south along a coastline—marine life varies by how often it is exposed to air and scalding sun, or how deeply it is covered by water.

The first problem Gilman and Sagarin encountered was finding which way Hewatt's line ran. The location of the first bolt was obvious, but finding the second proved difficult. When they finally did find it, the line ran in a different direction than they had initially thought. Armed with the location of both bolts, they meticulously constructed a map of squares in between, precisely repeating what Hewatt had done. In all, they counted 125,590 animals, representing 136 different species.

Once the counting began, Gilman and Sagarin quickly saw dramatic changes in the makeup of the intertidal animal and plant community. What they found was a substantial species shift in the intervening 60 years that was best explained as a northward migration of species away from warming water—exactly the kind of response predicted by models of human-caused climate change.

When Hewatt first looked at the tidepools, there were no tube snails or sunburst anemones at all, although both were prevalent further south. When Gilman and Sagarin did the same, they found hundreds of the southern anemones, and as many as 229 tube snails crowded into 1 square yard. In all, ten of eleven species previously identified as southerners increased significantly. Six of eight northern species decreased significantly. Those changes showed up regardless of what the animals ate, how they reproduced, or where they sat in the taxonomic hierarchy. Meanwhile, daily temperature records showed waters there had warmed about 1.8°F since Hewatt squatted in the surf.

They looked at all the masses of data in front of them, and considered different reasons for the changes. The best they found was the increase in temperature. The combination of counting so many species and having such a rich set of temperature data going so far back means that their findings withstood substantial statistical scrutiny. In other words, it was almost impossible that what they found could be simply chance. "If you had a table with a diagonal line and some marbles," Sagarin says, "and you just dropped them randomly, there's no way you'd ever get the pattern we saw."

Gilman did some separate work looking at one seaweed species growing on the granite. Using a study from the 1960s by Peter Glynn, now a researcher at the University of Florida, Gilman did a similar rock-by-rock comparison of turfy red algae called *Endocladia muricata*.

"The areas where Glynn worked just became deserts, in a sense, because there was no algae there," she explains. "But if you looked a foot lower, most of what used to be up in his research area shifted a little bit lower. We think that has something to do with temperature changes, because as it warms up, everything at low tide becomes a little warmer and dryer. So it sort of makes sense that things would tend to shift a little bit lower so they're exposed to air for shorter periods of time."

The snails, sea stars, and anemones Gilman and Sagarin counted were all small. But the shifts they recorded—and the implications of those changes—were anything but.

These were "massive community reorganizations," Sagarin says. And the implication of such a finding is that "it seems that species respond to climate change in the present day. It's not just going to happen at some point in the future."

"It was one of the first studies to suggest that you could go out and find [climate-related] changes already," Gilman adds. "But it was completely unexpected to me. I thought I was just counting snails for a summer project."

The results have held up over time. Sagarin, who received his Ph.D. from the University of California at Santa Barbara in 2001, has resampled the Hewatt line regularly. And while he has not found new changes with the same kind of drama as those he and Gilman found in 1993, he also has not found anything to indicate that the species are shifting back.

"Changes in the short term—every few years—are much smaller than the dramatic, aggregate changes that we saw when we looked 60 years later," he says. "So it just suggests that, even though each of these organisms is changing in its own time frame, real ecosystem evolution takes a long time."

Baxter points out that what his students found in the intertidal area by Hopkins are the same kinds of community changes one observes climbing up into the Sierra Nevada mountains. And if climate is affecting the globe, and therefore the Sierras, the same kinds of changes are happening there on a different scale. Scientists believe it takes something on the order of 300 years to see the tree line shift; with *Endocladia*, the timeline is probably about 10 years. That, he says, makes studying intertidal areas ideal for observing the ways a community of plants and animals responds to a warming world.

"This shows that, at a level we wouldn't have expected, the animals at Hopkins have reacted and responded to a small change in climate by totally restructuring their community," he says. "Organisms are more susceptible to small changes than we ever would have predicted. I think that it is clear that we are in the middle of a great experiment."

## Nobody Noticed

The findings at Hopkins were remarkable in themselves, but also significant because the changes had been under way for decades and no one noticed. The first sunburst anemones appeared in 1947. Seaweeds that once covered the rocks—not just Gilman's *Endocladia*—completely disappeared. "The astounding thing is that we didn't know it was going on," Baxter says with a growl. "This is a marine biological station!"

If not for Baxter's tenure and tenacity, the changes might still be going unnoticed. Those slow shifts are easy to miss or to write off as a temporary change. Without a basis for comparison or long personal experience, population shifts in small ecosystems are nearly invisible.

The best way to avoid missing those changes is to track small details over long periods. But in the world of scientific research, there is a simple fact: Monitoring—the kind of meticulous counting that Sagarin and Gilman did, repeated over time—simply is not glamorous. It is not quick, it generally does not attract money, and it does not interest top researchers who can more easily find funding and fame with other kinds of studies. Even dramatic findings such as those at Hopkins and elsewhere do not often inspire funding agencies such as the National Science Foundation.

But when trying to understand how a warming climate evolves—and the effects of such warming on all natural processes—monitoring is the best and possibly the only way to document reactions to a phenomenon that every year becomes more obvious and less theoretical. Researchers examining what long-term data are available have uncovered alarming and portentous evidence of a changing world.

One area where monitoring has been funded is in the southern portion of the California Current, a 600-mile-wide swath of south-ward-flowing water running along the western U.S. coastline, roughly from Oregon down through California. Part of a great subtropical gyre of currents in the northern Pacific Ocean, the California Current is the eastern end of a swirling seawater highway that circles from Japan to Oregon, south past California just into Mexico, across to the Philippines, and back up to Japan to begin again.

Along the way the water changes. As it crosses the Pacific toward North America, more water comes in through rain than leaves through evaporation, so the overall current becomes less salty. As it comes down the U.S. coastline, it meets cold, salty water heading north on another current, mixing in great meanders and eddies that can be up to 300 miles wide. The current turns west again south of California to start the process again.

In 1949, a combination of state and federal organizations began monitoring physical, chemical, biological, and meteorological facets of the California Current under the auspices of the California Cooperative Oceanic and Fisheries Investigations program, known as CalCOFI. It was designed in part to track many factors affecting commercially important fish species such as mackerel and sardines. The data gathered under CalCOFI include air temperatures, wind speeds, nutrient levels, salinity, water temperature on the surface and deep below the surface, and the abundance of larval fish and zooplankton—the smallest marine animals.

The early monitoring cruises brought researchers as far north as the Oregon border, but the surveys were scaled down to meet budget demands in 1970. But those first 20 years of data were enough to show that what happens in the south and what happens in the north tend to be the same. The data since 1970 cover the area between San Diego and Santa Barbara. It is the largest, longest-term data set of its kind on the West Coast.

What happens if you watch those data change over the years? Your findings might echo those of John McGowan, an oceanography professor at the Scripps Institution of Oceanography in San Diego: As water temperatures have risen, the base of the marine food chain off the coast of California has crashed. And one by one, the fish and birds farther up that food chain are crashing, too.

Life in the ocean begins with tiny plants known as phytoplankton. Like all plants, phytoplankton need light to drive photosynthesis and nutrients to feed the process. Although it is somewhat counterintuitive, the richest and most nutritive ocean waters are the coldest and heaviest. Strong winds do the work of stirring the system and pulling the nutrient-rich waters up toward the light.

The first problems showed up in conjunction with El Niños, short-term changes in ocean temperatures that tend to increase the warm water along the western U.S. coastline, reducing the food that boosts the phytoplankton. But researchers like McGowan noticed a difference between early El Niños and the later ones. Numbers of zooplankton—the tiniest animals in the food chain, which depend on the phytoplankton—dropped during the El Niño of 1957 to 1959 and then quickly rebounded. But after subsequent El Niños during the 1983 to 1984 and 1997 to 1998 seasons, the zooplankton did not come back.

In 1995, going back through the accumulated years of data, McGowan reported a staggering finding in *Science*: Zooplankton numbers in the California Current had dropped by 70 percent. The CalCOFI data show a sharp increase in California Current water temperatures in 1977—at the same time the zooplankton numbers crashed.

"It's the largest change ever measured in plankton productivity in the ocean," McGowan says. "This enormous change in the zooplankton in the California Current could not be detected from year to year. It took several decades before we discovered this big drop, by at least 70 percent or even up to 80 percent."

If you pull out the bottom stone in a pyramid, you expect the structure to come tumbling down. With that huge loss at the base of the food chain, reverberations throughout the system that depended on it were inevitable. Since McGowan's study came out, declines of species throughout the area have been attributed to the loss of zooplankton and the warming water.

The crash showed up in fish, although it is often tough to tell if such declines come from too many nets or too little fish food. But even when researchers look at species for which human markets have no appetite, they find precipitous declines. The larvae of *Leuroglossus stilbius*—a fish of so little market value that it does not even have a name in English—historically are the third most abundant in the California Current. Counts of its larvae dropped 50 percent after 1977. Another similarly ignored species with no common name, *Stenobranchus leucopsarus,* saw its larvae drop 42 percent after the sharp temperature rise. Its larvae are typically the sixth most abundant in those waters.

In 1967, aerial surveys found 70 square miles of kelp forests along the long California coastline. In 1989, that number dropped 42 percent. By 1999, the most recent year for which data are available, the total plummeted to just 17.8 square miles, down 75 percent from the 1967 survey.

## Breaking the Top of the Chain

But the most dramatic decline came to the sooty shearwater, a predatory seabird at the top of the marine food chain.

"In the 1960s and 1970s they were present in the tens of millions," McGowan says, "the largest population of pelagic [marine] seabirds in the entire California Current. They dominated it. Millions and millions of them." The birds feed on juvenile fish and larger zooplankton. Researchers began looking at the birds regularly in 1987. By the 1990s, the population of sooty shearwaters—like the guillemots in Alaska—had crashed, with numbers down 90 percent.

"The decline of the sooty shearwater is very dramatic," McGowan says. "And that clearly is not due to man harvesting it, or at least not directly."

As of 2003, water temperatures in the California Current are back down to their long-term average, but zooplankton numbers have not changed. In fact, McGowan says, samples taken in February 2003 showed the lowest abundance ever recorded.

Eight years after first reporting the zooplankton die-off, McGowan thinks he knows why it occurred: The warm surface layer of the ocean got so deep that the nutrient-rich waters below could not get close enough to the light to help the phytoplankton grow.

McGowan compares that oceanic phenomenon to the familiar warm layer on a lake or swimming pool. When you go in, he explains, "Sometimes your feet are cold and your upper body's warm. Warm water floats on top of cold because it's less dense. And because there's this sharp layer between the warm upper water and the cold lower water, it makes it difficult to stir."

In the California Current, McGowan and his colleagues have found that the line between warm surface water and cooler, nutrient-rich water became substantially deeper in 1977, when everything heated up. The deeper that line goes, the harder winds need to blow to mix up the nutrients. And the winds have not changed. That means the basic driver of ocean life is on an extremely low setting. "Less productivity, less plankton and birds and squid and fish," he says.

Although McGowan's work is limited to southern California, the earliest CalCOFI data showed that what happened on the south end of the California Current was very similar to what happened further north. That means the die-offs he has documented could be repeated all the way up the West Coast.

"The whole system pumps up and down and up and down," he says. "It's one system, in spite of all of the eddies, in spite of all the meanders, in spite of all the species that are involved. We know that the temperature change is still synchronous, all up and down the coast. That has been extensively tested."

And that is perhaps the most important implication of McGowan's research: The ecosystem crash documented in southern California could happen along with warming in any part of any ocean anywhere. Some British studies are turning up evidence of deepening surface layers of warm, nutrient-poor water, McGowan says. U.S. agencies and the Intergovernmental Panel on Climate Change have documented higher ocean surface temperatures around the world.

"This 1977 regime shift that we thought was something that happened in the North Pacific, there's evidence globally that temperatures took a big jump up," McGowan says. "It wasn't just California, the West Coast, the North Pacific or the Gulf of Alaska. It was everywhere."

"It's a very, very serious problem," he adds. "And the seriousness comes from the fact that we really don't know what the consequences will be." The crashes McGowan has seen "could be a generalized response of the whole ocean—all the world's oceans—to heating. It might look something like what we see in the California Current."

As with Gilman and Sagarin's research, McGowan cannot definitively prove that increased carbon in the atmosphere caused heating that was the direct and sole cause of the changes he has seen. But his research continues to suggest temperature increases as the most likely driver for his observations.

## Rising Waters

Climate change appears to be making big-picture changes to transcontinental currents and the rocky interface of shore and ocean, but rising sea levels also combine with runaway development to devastate local environments. In the last remaining salt marshes ringing San Francisco Bay, a bird and a secretive mouse are steeling to fight their own battles with climate change. The problem there is not that the water is getting too warm, but that it is getting too high.

As glaciers melt and oceans rise, so too will the waters of the famous bay. That would not necessarily be a problem, since the marshland plants and animals in theory could simply move upslope, in much the same way Sarah Gilman's seaweed moved down toward cooler water. But two endangered species—the California clapper rail, a secretive bird that does not much like to fly, and the salt marsh harvest mouse, which lives without drinking fresh water—are stuck between rising water and asphalt, the latter covered by multimillion-dollar development. If or when the water rises, they will not be able to afford the new rents.

Marge Kolar, who manages the Don Edwards San Francisco Bay National Wildlife Refuge, says that 150 years ago the water was ringed by 50,000 acres of muddy tidal flats and 190,000 acres of lush tidal marsh. Today only three-fifths of the tidal flats remain, and a mere one-fifth of the marshland. The clapper rail and harvest mouse live nowhere else on earth, and need the marshes to live, hide from predators, and feed. The clapper rail was one of the first birds put on the federal list of endangered species, and only about six hundred individual birds are still alive to perpetuate the species. No one knows how many of the mice there are. Their population is assumed to have declined as much as their habitat has: 79 percent.

From the hill above the refuge's visitor center, the marshes look like geometric farm plots outlined by sandy dirt roads. Here, however, the fields are ponds used by Cargill Salt to evaporate water and harvest salt, and the roads are earthen levees 5 to 6 feet tall that keep the waters of the bay out.

Before those walls went up, cordgrass grew out of the mud ringing the bay at low tide. As it eased away from the water's edge, the cordgrass gave way to salty, bitter pickleweed, whose round, segmented stems grow like anemone arms, historically feeding the native Ohlone Indians and still the major source of water for the harvest mouse. Further above the tide line, the pickleweed in turn gives way to the shrubby, yellow-flowered gumplant that the clapper rail runs to when it is hiding from predators.

It is possible to reverse time and turn salt ponds back into marshland, Kolar says. It just takes breaching the levees and letting the bay waters go to work. As an example, she points just east of the refuge's visitor center. The refuge bought the land from Cargill and, in 1985, breached the levees, and let tide water flow back in. Although it did not have the complexity of the more mature marsh nearby, by 2000 harvest mice were in it, munching the new pickleweed and making themselves at home.

But the busy commuter street just behind the recovering marsh, Thornton Road, is a sharp dividing line between the ecological reality of the harvest mouse and the economic reality of the area's famed computer industry. Just behind it rise the shiny offices of Sun Microsystems, which paid Cargill $477,000 per acre for the old salt pond at the peak of the Silicon Valley boom. Kolar is quick to say that the refuge did not pay anything like that amount of money when it bought the land for restoration.

In the spring of 2003, Kolar's refuge and the state of California joined forces to buy another 16,500 acres of salt pond from Cargill at bargain, nondevelopment prices in a bad economy—slightly above $6,000 per acre. The purchase nearly doubles the total protected area around the bay, bringing it up to 38,500 acres of total refuge land. Kolar says the U.S. Fish and Wildlife Service and two state agencies are beginning to plan for restoration, and they will take rising water into account.

Nevertheless, the purchase became controversial because details about pollution and the true value of the land were kept from the public. Newspaper reports say the $100 million deal was based on outdated economic assumptions—especially prickly given the collapse of the Silicon Valley economy—costing taxpayers unnecessary millions.

And while it may be physically easy to turn salt ponds back into marsh, factors other than development and rising sea levels are complicating restoration. Especially in the south part of the bay, near San Jose, the land ringing the bay is slowly subsiding. Between the dropping land and the rising water, the native endangered species are getting further squeezed. For them, the simple solution is to break the salt-pond levees. But doing so now carries another political cost: Those levees currently protect the city of San Jose from encroaching bay water.

The new land purchase will do the clapper rail and the harvest mouse little good if it is inundated and there is nothing but asphalt when the cordgrass, pickleweed, gumplant, and assorted species that depend on them try to move upward away from the rising water.

But exactly what is happening on the edges of the bay is still murky. At the same time as the water is rising and the land subsiding, development and erosion are constantly spilling more sediment into the bay. Those sediments fill in the very areas where the mouse and the rail live, with the possible effect of creating a new ring of habitat.

The big question is which will come first: rising water pushing these endangered creatures onto asphalt or rising sediments building them new homes. "Will [sediment deposition] keep up with sea-level rise?" Kolar asks. "That's a question people haven't answered yet."

In the world of science, nothing is ever proven. Researchers say their findings "show" a certain thing, and may discredit some previously held notions. McGowan, Sagarin, Gilman, and Baxter cannot say that the crashing food chain or the migration of intertidal species they documented is "caused" by a warming world, although they all seem to believe that is the case. But all the pieces together begin to form a picture. If it is not a picture of global warming, it is certainly a picture of what global warming could bring.

Chapter Eight

# Australia, Florida, and Fiji: Reefs at Risk

*David Helvarg*

I first heard about coral bleaching from Billy Causey, the manager of the Florida Keys National Marine Sanctuary. We were sitting in his office deep in a 67-acre hardwood hammock on Marathon Key. It is a place where ospreys, egrets, cormorants, fat black snakes, hermit crabs, parrot fish, even an old tropical fish collector like Billy can still find refuge from the Kmart mall sprawl out on Route 1. Thickset with iron-gray hair and sea-gray eyes, Causey, who moved to the Keys in 1973, sounds like some Old Testament Jeremiah as he recalls the gradual decline of the reef during the years he's been here.

Unfortunately, while among the most diverse of marine habitats, the world's massive coral colonies are also fragile structures, living within a narrow range of clarity, salinity, low-nutrient chemistry, and temperature.

Fig 9: Minden Pictures/ Fred Bavendam

"Throughout the '70s we saw various problems but constantly clear waters with typical hundred-foot visibility," Causey recalls.

"In 1979, we had a warm water spell and big vase sponges started dying," he continues. "In June of 1980 we had a pattern of slick calm weather and thousands of fish were killed. This was the first signal to me that things were tilting the wrong way. Then in 1983, with an explosion of onshore development, there was an urchin die-off. In 1984, there was another doldrums and the reefs bleached down to Key West. Maybe five percent of the coral died. In May of 1986, when we had hardly seen black band disease [characterized by dark bands of dead coral on otherwise healthy specimens], I went out to take a picture of it. I saw four-dozen massive outbreaks within an area about 400 feet in length."

Causey pauses to listen to a passing bird cawing over the still, aquamarine waters of the Gulf a few yards away. Further north I have noticed that the fringing waters of Key Largo have taken on a greenish lime Jello hue.

"In June of 1987 we got a slick calm," he continues. "On July 13 we went out and saw all the corals turning mustard yellow. Then they went stark white. Then we began getting reports of similar bleaching in the Caribbean and on the Indo/Pacific reefs and we realized something global was going on. We began looking at this as the canary in the coal mine. Meanwhile, the National Oceanic and Atmospheric Administration—NOAA—was reporting 1987 as the hottest year on record and the 1980s as the hottest decade." These records would all fall in the 1990s.

The bad news multiplied, Causey says. "In 1990, we had the first big losses linked to bleaching where the coral didn't come back. We lost most of our fire coral that year. There was another benchmark year in 1997, with coral bleaching all around the Caribbean. Lots of living coral just went away in 1998, a catastrophic bleaching event. But remote reefs in the Pacific were also being lost, so it gave me a sense that this wasn't an isolated event—the result of our failure to act. There were back-to-back severe bleaching in 1997 and 1998, then Hurricane George hit."

Causey shakes his head, as if unwilling to believe his own unremittingly bleak narrative. "You look at old photos and film of the reef and you realize what was lost," he says. "If you were lucky enough to be here 20 or 30 years ago, you know."

I do. As a teen, I snorkeled through Keys' waters so clear and vibrant with exotic life and color they were almost scary.

After talking with Causey, I will do some diving in the Keys, including a couple of dives down to Aquarius, the world's last underwater research habitat, 7 miles off Key Largo. The habitat is a 48-foot cylindrical structure resting on four steel legs planted on the bottom in 60 feet of water. Its yellow body is rusting in spots and encrusted with weedy growths being grazed by roving schools of fish. A couple of large tarpon in the 100- to 150-pound class circle it curiously, shadowing me as I swim under the metal skirt of the habitat, popping up in the wet room where a school of yellowtail snapper huddle discreetly at the edge of the entry pool. Beyond the wet room there is a lab, shower and toilet, kitchen and berthing area with two sets of triple bunks. Scientists, living here for up to 8 days at a time, have a unique opportunity to study the world's third-largest reef system.

Unfortunately the reef they are studying is also dying. Where once branching corals grew, I find only skeletal sticks in faded rubble fields. Many of the abundant rock corals are being eaten away by diseases that have spread in an epidemic wave throughout the Keys. The names of the diseases tell the story: black band, white band, white plague, and aspergillus, a fungus normally found in agricultural soils that can shred fan corals like moths shred Irish lace. The corals are also being smothered under sediment and algal growth linked to polluted run-off and are periodically bleaching white as a result of warming ocean temperatures.

## Australia's Coral Crisis

I did not see some of the worst bleaching effects in the Keys, but 8 months later I get an opportunity to investigate coral bleaching and its links to climate change as part of a magazine assignment in Australia and Fiji. It will turn out to be a challenging job full of too much travel, miscommunications, political conflict, a dive in which my air cuts off, and another in which I cut my hand on coral, resulting in the need for hand surgery. Still, given my paradise-rich itinerary, I receive little sympathy from my friends.

The first leg of the journey takes me from Washington, D.C., to Los

Angeles and on to Sydney where, during a 7-hour layover, I call an Australian journalist friend who gives me the name of one of his contacts, Jeremy Tager, head of the North Queensland Conservation Council. I catch the next flight heading up to Cairns in Australia's "Deep North," a land of rain forest and reef that juts up into the soft underbelly of Papau New Guinea.

In Cairns I rent a car and drive south on the two-lane Bruce Highway heading toward Townsville halfway down the length of the 1,200-mile-long Great Barrier Reef, passing through verdant green valleys and fields of sugar cane. Thirty hours into my trip I arrive at Mission Beach, putting up at The Horizon, a lovely spot with a wooden pool deck overlooking Dunk Island. I have a drink at their outdoor bar, trying to ignore Georgina, the mean-looking pheasant sitting on the chair next to me. I drive down to the beach for dinner and spot a 3-foot wallaby hopping through my headlight beams on the way back.

That night I finally get some sleep, despite the surrounding jungle sounds, which include the scuttle of large lizards and a loud white cockatoo screeching in the tree next to my bungalow. In the morning I hike down to an eroded beach with a river outlet and a wooden sign: "Warning: Esturine crocodiles inhabit this river." As naturalist author Edward Abbey said, "If there's not something bigger and meaner than you are out there, it's not really a wilderness."

A few hours later, I pull into Townsville, a waterfront city of 120,000 that bills itself as the capital of the Great Barrier Reef (even though Cairns took away most of its tourists years ago when it opened its big airport). Townsville is trying to make a comeback with a new IMAX theater, reef aquarium, and casino, but also hedging its bet with a zinc-processing plant, army base, and government offices including the Barrier Reef Park Authority and Australia Institute of Marine Sciences (AIMS). I call Tager, who then invites me over to his island home, a half-hour ferry ride from Townsville. He greets me on the dock, where he quickly fulfills his duties as a professional activist by briefing me on some of the environmental issues that threaten the world's greatest marine park.

"There's bottom trawling by commercial fishing boats still allowed in the park, also nutrient run-off from the sugar industry, forest clearing for cattle and new developments along the coast," Tager says. "There's talk of a gas pipeline, plus the oil industry keeps pushing to open up offshore drilling just beyond the eastern boundary of the park in the Coral Sea. The government, of course, keeps pushing these schemes while failing to pursue renewable energy strategies."

Magnetic Island is like Catalina Island with death adders. "They like to lie under piles of leaves," Ann Trager advises, as she shows me the family's wooded, leafy backyard. The most visited of the Great Barrier Reef islands, Magnetic is rich in tropical birds and wildlife. It is home to adders, pythons, rock wallabies, sea eagles, flying foxes, curlews, koalas, and equally cute marsupial possums. It also has its own living coral reefs, but no longer in Nelly Bay.

The hottest year in recorded history, 1998, saw a global outbreak of coral bleaching as corals' thermal tolerances were exceeded by a combination of gradually warming sea temperatures spiked by that year's El Niño phenomena. The idea that climate change is accelerating El Niño warming and the La Niña ocean cooling that follows it has become a subject of growing scientific concern. The U.S. State Department's Coral Reef Task Force reported that the extensive 1998 global bleachings were a direct result of climate change caused by the burning of fossil fuels.

"When we saw 1,000-year-old coral colonies bleaching and dying, that's something new, at least in recorded human history," agrees Paul Hough, a friendly, sun-reddened Magnetic Island resident and research scientist with the Great Barrier Reef Marine Park Authority.

Hough specializes in coral reproduction. "I was looking at the corals that didn't die, and found their reproduction was down to 40 percent of normal the first year after the bleaching and was at 80 percent the second year," he says. "Now they're experiencing 1.5-degree below-normal water temperature from La Niña, so we've had a four-degree swing in four years. With greater frequency and severity of El Niño/La Niña events, it will be more difficult for corals to recover from these kinds of impacts. I think we're seeing not a crash, but a slow decline of the [reef] system."

Hough and other observers believe nearshore fringing reefs are at the most immediate risk because they already face other stresses from human development, run-off and cyclones (or hurricanes as they are known in the Florida Keys).

I am visiting Magnetic Island following the two wettest months in North Queensland's history. Big gum trees and foliage are still down from Cyclone Tessa that struck 2 weeks earlier.

My second evening in town I'm invited to the Magnetic Island Film Festival where the friendly, beer-drinking crowd of more than one hundred is mostly costumed, in bathing suits or in drag. Here I learn that the island's two thousand residents are divided over a proposal to revive a failed harbor project. The real-estate development would add six hundred new housing units to the community and its over-stressed septic systems. To date, Queensland, like the Florida Keys, has not allowed climate-enhanced storms, increased rainfall, sea-level rise, or dying reefs to stand in the way of its beachfront development schemes.

The island's once-healthy Nelly Bay, a kind of neighborhood reef where kids first learned to snorkel, is one of Hough's seven study sites. "Unfortunately the cyclone following repeated bleachings was the final nail in its coffin," he tells me. I decide to visit it anyway, to see for myself.

I head down to the beach past stately banyan trees, hoop pines, coconut palms, and a sign reading: "Warning: Marine Stingers Are Dangerous . . . Emergency treatment [for] severe box jelly sting: Flood sting with vinegar. If breathing stops, give artificial respiration."

Luckily I have borrowed a stinger suit (what we Yanks call a rash guard or dive skin) from Tager. I swim out to the black buoy that marks Hough's research site and begin free diving. The bottom is a rubble field of broken branch corals, dead bleached and gray silt-covered hard corals, and a few small fish. A burrowing clam is encased in the limestone skeleton of a dead rock coral. Its blubbery mantle is striped and spotted with the blue, purple, and green colors of healthy symbiotic algaes, giving it the look of a fashion model posing in a cemetery.

The Australia Institute of Marine Sciences (AIMS) is located 40 minutes south of Townsville. To get to it, you drive out Cape Cleveland, a wetland and wilderness peninsula full of storks, egrets, cattle, and the occasional wallaby bouncing off into the brush. Eventually you arrive at a metal gate with a speakerphone in the middle of nowhere. The remote gate opens. On the other side of a grassy hill is AIMS' low-rise glass-and-concrete research complex, and just beyond it down a brushy path is 5 miles of wilderness beach facing out on the world's largest living reef.

"Climate impact has happened. The four most serious bleaching events were in 1987/1988, 1992, 1994, and 1998, which was the biggest," explains Katharina Fabricius, a bright, vivacious AIMS research scientist with firsthand knowledge, also a refugee from Germany's harsh winters. "Corals can take a fair amount of disturbance, they're not fragile," she tells me in an office filled with soft coral samples as varied and unique as snowflakes. "If these disturbances become more frequent, however, weedy species will take over. You already see branching species replacing massive slow-growing brain corals. We lost a 1,000-year-old coral head off Pandora reef in 1998. These reefs are really the canaries in the coal mine where you now see a whole ecosystem being impacted."

I tell her I know an Antarctic scientist who thinks his penguins are the canaries for climate change.

"Ten years ago people were blasé about this being a pristine area," Fabricius continues. "Now with climate change, even the most conservative projections are pretty bleak. And if your [Australian] government wants to sell brown coal, they may not be likely to consider alternative fuels or solar or other changes that need to take place."

In fact, Tager and other local environmentalists are now fighting a plan to start mining shale oil in the rain forests of North Queensland, arguing that the last thing the world needs are new sources of fossil fuels. Still, not everyone is convinced.

"In many ways the jury's still out on the global climate effect on coral bleaching," claims Virginia Chadwick, a former regional tourism minister and the political appointee who chairs the Great Barrier Reef Marine Park Authority. "Not to say this apparent correlation between bleaching and temperature isn't a worrying trend," she adds. "From a local management agency point of view we're wondering about adaptability, about corals' ability to adapt to temperature changes."

"We have no evidence corals can adjust to rapid temperature changes," counters Fabricius. "Maybe they can over hundreds of thousands of years, but that's not the scale we're now dealing with."

Having seen dead coral, I decide to take a diving trip to Kelso, one of the outer reefs that has recovered from the little bleaching it did suffer in 1998. It is nice to be in a living aquarium again, with big coral walls, bommies (coral heads), and canyons littered with fish; luminescent chromies, coral trout, bat fish, rainbow-hued parrots munching coral and pooting sand; and the usual clown fish hiding in the arms of purple-tipped anemones, waiting for someone to take their picture. I spot a giant clam about 6 feet across, looking like a living boulder, also a big lobster hiding under a shelf. There are purple sea stars and cushionlike sea cucumbers, trigger, trumpet, red emperor, and unicorn fish, and lots of bright juvenile fry hugging the reef for protection. A black-tip reef shark cruises past, adding a dash of predatory grace to the mystery and magic of this healthy reef.

"Ours is a large reef region, more robust than the Florida Keys or the Caribbean with more than 420 species of corals, six to seven times as diverse as your Atlantic reefs," Hough explains to me.

"What does that mean in terms of long-term projections?" I wonder.

"Larger diverse communities [like the Great Barrier Reef] will last longer," he says, "but North America is going to get hammered."

Not a happy thought for Billy Causey or anyone else who loves the Keys.

## Sinking Fiji

From Sydney, I catch a flight to Nadi on Fiji's main island of Viti Levu. There I catch a second cross-island flight aboard a Twin-Otter prop plane. There are six of us taking the night flight, including two ladies in saris, as we fly over mountaintop clouds illuminated by star-spangled southern constellations.

In 1874, the kingdom of Fiji, known for its fierce cannibal warriors, became a British Crown colony. The following year a measles epidemic killed a quarter of the native Fijians. Sympathetic to their suffering, Sir Arthur Gordon, the first colonial governor, forbade Fijians from selling their lands or working as day laborers. British sugar planters got around these restrictions by leasing land from the Fijians and bringing in Indian laborers to work as indentured servants. When this feudal system finally ended in 1916, forty thousand Indians opted to stay on. Today Indo-Fijians make up almost half of Fiji's population of 750,000. They are part of what the tourist brochures call a "vibrant, multicultural society," but one I quickly come to recognize as yet another society fractured along racial, ethnic, or religious lines.

I share a ride into town with an American named Steve, who says he is with the State Department but reminds me of some of the CIA types I used to know in Central America. He has just visited Somoa, Palau, Tuvalu, and the Marshalls. "It's damn scary," he says. "Some of those places are only a few meters above the water. You drive down the main road and you're looking at ocean to the right and left. If a storm comes you find a big palm tree to hang onto and hope you don't get washed away. The State Department's really not paying enough attention to these environmental threats."

Suva itself is a typical ramshackle Third World capital, with a central market and bus depot, a university, Japanese fishing vessels tied up at its docks, and foreign-owned clothing plants on its outskirts. It has little charm and few of the resort amenities that attract tourists to Nadi and the outer islands.

"The political climate is heating up and heading towards a real frenzy," a local reporter tells me after I have settled in at the Centra, the harbor-front hotel on Victoria Parade where foreign business types and journalists stay. This week it is also hosting a meeting of the Fiji Human Rights Commission.

The day I arrive, Prime Minister Mahendra Chaudhry overrules his secretary of home affairs and decides to permit a protest march by Fijian nationalists opposed to Indians holding political power. Steve later tells me the U.S. ambassador helped convince the prime minister that this was the right and democratic thing to do. Chaudhry is the country's first elected Indian head of state since a Fijian nationalist coup in 1987. But while ethnic tensions seethe, it is climate change that represents the greatest threat to Fiji's future.

Native Fijians tend to work in the tourism industry, while Indians make up most of the 23,000 families raising sugar. Both of these primary industries are threatened by global warming.

Fiji's director of environment, Epeli Nasome, is dressed conservatively in a white short-sleeve shirt and brown business skirt or *sulu vakataga*. He sports a neat graying mustache, wears glasses, and has the cautious manner of bureaucrats throughout the world, which makes what he has to say all the more disturbing.

"We can feel a change already in our weather system here, with longer droughts that impact our western division [on the main island]," Nasome tells me. "We're having more rain, more rainy seasons with higher rainfall and flooding has spread to the western side of the island, which is normally known as the dryer part. And this coral bleaching is a new area of concern, and is on a more extensive scale than we've ever seen before."

The manager of the Jean-Michel Cousteau dive resort on the island of Vanua Levu, the second-largest of Fiji's more than three hundred islands, recently began complaining about the state of the road into the upscale retreat. He claims it is in such bad shape visitors may not want to return to Fiji (if the daily newspaper reports of arms smuggling and a possible coup do not scare them off first).

While promising to improve the road, the public works minister points out that this has been one of the rainiest years ever, and bad for roads throughout the islands. Two weeks earlier, Labasa, the main town on Vanua Levu, flooded. A year ago the cities of Nadi, Ba, and Lautoka in the western sugar-growing region of Viti Levu also suffered major flooding. This followed an 8-month drought that devastated the sugar industry.

"Seasonal shifts are becoming more extreme," Janita Pahalad, the manager of climate services for the Fiji Meteorological Service, tells me. "Another problem is that with global warming, night-time temperatures are increasing, but the sugar industry needs low night-time temperatures to increase the sucrose content of the cane."

Pahalad has written a report suggesting that rising sea levels from global warming are also leading to increased salt water intrusion into water tables, changing the pH level of low-lying sugar fields, which can drastically reduce their productivity.

On Fiji's low-lying islands salt water intrusion can come from above as well as below. On Moturki, a small island not far off eastern Viti Levu, a 1999 storm surge crossed over the island, ruining the islanders' crops and salting their freshwater. The nine hundred residents had to have freshwater shipped in for the next 8 months.

"In a village in [the northern island of] Laucala, they're complaining about drinking seawater," explains Robert Matau, senior sub-editor at the *Fiji Post*. "The whole [western] Yasawa island group has a huge [salt water intrusion] problem." Matau is a large bearded journalist, with a traditional *sulu* skirt, round face, and skeptical brown eyes. "Newsrooms neglect the environmental story," he complains. "What made global warming real for me was when I returned to Kadavu [an island group south of Viti Levu] to pay my respects to my great-grandfather in 1992 and found his grave half in the ocean. Then in 1997 a cyclone and tidal wave washed away the road, jetty and much of the area's shoreline. Now in my mother's village of Muani they're moving houses inland and trying to build a coral seawall. Two weeks ago the island's main village of Tavuki flooded. If you're on an island with no [mountain] slope I believe you seriously have to start thinking about moving."

Even on high green volcanic islands such as Taveuni, Fiji's third-largest, where they grow copra, cava, and taro, most of the fourteen thousand residents live along the shore. Leigh-Anne Buliruarua is from the Taveuni farming village of Vuna. "In the last 30 to 40 years they've made a rock seawall around my village. It's all we can do, we can't afford a regular seawall," she tells me.

"We have to assess the vulnerability of our coastal areas and then we can see how we can adapt with sea walls or relocation of settlements and people," Environmental Director Nasome worries.

Still, Fiji is not in as bad a state as the low-lying Pacific island nations of Kiribati, the Marshalls, and Tuvalu, which may literally be subsumed by rising sea levels linked to fossil-fuel-driven climate change. There are already diplomatic discussions under way about where to resettle the Island's environmental refugees. American attorneys have also journeyed to Tuvalu, eager to take on the island as a client in a lawsuit about global warming.

Such suits are also under discussion in Fiji. "If something is a survival issue, then it is a human right," Dr. Shaista Shameem, an attorney and director of the newly formed Fiji Human Rights Commission, tells me. She has been thinking about a possible suit against industrial nations and oil companies on behalf of those whose islands and cultures are being drowned. We talk in the Centra conference room where her group has been meeting, but after 4 days of interviewing government bureaucrats, activists, and academics (at the University of the South Pacific), I am ready for a different kind of in-depth reporting—from the Garden Island Resort on Taveuni.

I fly there via Vanua Levu over small islands and wild aqua-blue reef lines, with breaking waves stretching toward the horizon. "Imagine we had no reef. One big wave and we'd be gone," Ritesh Lal, my 21-year-old Taveuni taxi driver tells me on the way from the airstrip, when I mention that I am writing a story on reefs and climate. We are driving on the island's main dirt road that runs along the water's edge. "If you make the change from these fossil fuels maybe these oil guys won't be making enough money and that's it," Ritesh suggests with a grin, and then more seriously says, "I get behind these trucks and buses and they have all this black smoke because they don't maintain their engines. Or people clean their engines and let the water run down the drains. And you know all those drains end in the sea."

Lal tells me he got his environmental education in school. "And that's where I went to school," he says as we speed past the Methodist High School across from an eroding seawall. A few minutes later we pull into the hotel, a converted Travel Lodge now equipped for scuba diving in the Somosomo Strait.

During my first dive I notice that the reef looks like a snowstorm has passed over it. About a third of the corals have bleached white. Some of the staghorn and other branching corals are wedding cake white. The worst-hit area is Beqa Lagoon south of Viti Levu, one of the most popular dive sites in Fiji, which has suffered more than 80 percent potentially lethal bleaching as the a result of a huge pool of warm water first spotted by U.S. satellite.

"Did you see that bleaching?" I ask one of the other divers, a travel agent from Minnesota, after climbing back onto our boat. She looks at me curiously. "When the water heats up the coral polyps either lose or expel the algae that give them their color. Unfortunately the algae also provide 70 percent of their food, so they begin to slowly starve to death," I explain to her.

"Really? I thought they were supposed to be white like that," she says. It is what scientists call a shifting baseline. If you have never seen a healthy vibrant reef, you might not recognize a dying one, even as you swim through it.

Still, much of the strait's reefs remain intact and otherworldly. We do a fast-moving drift dive over a reef swarming with small purple and orange fish, also triggers, groupers, and unicorn fish, some pretty Morish idols, and a Picasso fish (to see it is to understand the name).

We next dive a spot called Jerry's Jellies. The soft corals here look like green bushes, black spiders, yellow mushrooms, and red velvet cloth lain over rocks and also buttercups in purple, mauve, and violet. There is a sand field of garden eels waving like prairie grass in the current. Bigger fish are also lolling in the current waiting for the small fry to move from their cover among the brain and lettuce corals. I spot a parrot fish with a small rainbow-colored fish held in its mouth. As he spits it back out, I drift into some sharp coral, cutting my hand. At this depth the blood looks black as it spurts slowly in little ribbons that dissipate in the current. Back onboard the boat, I bleed a bit, hold my hand above my heart until the bleeding slows, and then bandage it. We take a break from diving on a beach in Vanua Levu, where we snack on fresh pineapple and homemade cookies. From here the green slopes of Taveuni remind me of the big island of Hawaii. A large fruit bat flies across the open strait, shards of nautilus shell crunch underfoot.

On our last dive I spot some silly-looking eels, blue with yellow pronged snouts that attract smaller fish. I also spot black-tip reef sharks cruising along the edge of the reef and two lion fish in a cavern, their fanlike quills both beautiful and tipped with poison.

Back on Taveuni, I visit the Wairiki Catholic Mission, walking down a dirt road past crumbling seawalls and a few uprooted coconut palms to a shallow rocky beach that has eroded away in recent years. A group of young Fijian kids are playing naked in the water. When they spot me taking photographs, they come over to pose and shyly tell me their names and ages.

Four days after I leave Fiji, the protest march that the U.S. ambassador had promoted as an example of democracy takes place in Suva. A group of armed men leave the march and take Prime Minister

Chaudhry and his cabinet hostage. This sets off a series of riots, killings, and a military coup. The tourist economy falters. As a journalist, I am naturally disturbed that I have missed the action, coming home with photographs of eroding seawalls rather than Fijians burning Indian shops. On the other hand, I know that while ethnic conflicts can be resolved, there is little the Fijians themselves can do to protect their kids from a future of rising seas and dying reefs.

The U.S. State Department estimates the value of coral reefs at well over $100 billion. The Florida Keys generate $2.5 billion a year from tourism. The Barrier Reef brings $2 billion a year (in U.S. dollars) in tourist revenue to Queensland, and Fiji's $250 million ocean-based tourism industry is the largest in that small country.

Other reef products and services include commercial fisheries, storm barrier protection for islands and coastlines, and potential pharmaceutical medicines including new cancer-fighting compounds derived from soft corals.

More than their utilitarian value, however, the reefs represent one of the most biologically diverse communities of life on earth, with more than 25,000 known species of plants and animals.

Moving weightless through their vast and intricate gardens, I find it hard not to feel a sense of moral obligation and outrage. If we let the world's reefs die, turning white as bone so that we can keep burning coal and oil, we will rightfully earn the scorn of future generations denied the awesome experience of our blue planet's living coral heart.

# CHAPTER NINE

# Pacific Northwest: The Incredible Shrinking Glaciers

*Sally Deneen*

It feels as if a giant meat locker has swung open, sending a cold, yet thin, wind blowing down South Cascade Glacier just outside North Cascades National Park in northern Washington. The sun glares. Everything is white. The expanse of snow acts like a big reflecting basin. Bob Krimmel, a scientist in a broad-brimmed hat and gloves, is initially winded by the altitude change, but spends much of the day trudging through brush to get to this spot—the longest-studied glacier in the northern Cascade mountains, the nation's most heavily glaciated area outside of Alaska. So much snow. And yet, the glacier is shrinking.

"It's very easy to see the glacier is much, much smaller," Krimmel says later, back at his office at U.S. Geological Survey (USGS). Seated at a computer, he looks at the side-by-side images—a photo taken in 1928 and another 60 years later. "In the last century, it's retreated about 1.2 miles," says Krimmel, a research hydrologist and the glacier's leading researcher. "Right now, it's about 1.5 miles long. It's lost about half of its length and half its volume."

South Cascade Glacier has become the poster child for global climate change in the Pacific Northwest, contends Jon Riedel, glacier researcher for North Cascades National Park. It is thinning so much, Riedel points out, that between 1953 and 2000 it lost the equivalent of 72 feet of water in thickness off its surface. That is about as tall as seven basketball hoops stacked on top of each other.

Call it the case of the incredible shrinking glacier. In this icy high country, forty-six of the forty-seven Cascade glaciers observed by Nichols College researcher Mauri Pelto were found to be retreating. Riedel, meanwhile, personally backpacks several miles to monitor four glaciers; he notices the lower-elevation, smaller glaciers on the west side of the Cascades are shrinking, a pattern also found farther south.

This melting promises to change the very image of the Pacific Northwest. Montana's Glacier National Park in 30 years may need to be renamed "the park formerly known as Glacier," as Seattle-based writer John C. Ryan puts it. A hundred of its 150 glaciers have vanished, and the pace is hastening. Or take Washington's white-capped Mount Rainier, that looming symbol of the Northwest depicted on Washington license plates and the label of a venerable local beer. The vast majority of Rainier's glaciers are receding, says Andrew Fountain, researcher and Portland State University geology professor.

"They don't recede because they're getting colder, you know what I'm saying?" Fountain says. Whatever the ultimate cause, he continues: "That's global climate change—right there."

In scene after scene played out around the Pacific Northwest, researchers are uncovering surprises that appear linked to the past century's average 1.5°F rise in temperature and, on the basis of what has happened so far, they predict serious problems to surface in the next century. The surprises are as varied as the region itself, from the dangerously delayed spawning of salmon in British Columbia to the practically regionwide shrunken snowpack and perhaps happier news—such as the discovery of a butterfly that has colonized Oregon and Washington from the south as temperatures warmed.

"We're seeing things that never happened before to our knowledge," says Elliott Norse, president of the Marine Conservation Biology Institute in Redmond, Washington. "These things are consistent with what we would expect in a world that is warming. It would be, in many cases, surprising if this weren't human-caused.

"When you think of the state of Washington, do you think of marlin or yellowfin tuna? No. Well, starting several years ago, people started catching marlin, which we think of as tropical and subtropical, and yellowfin tuna off the coast of the Pacific Northwest," says Norse, adding that while the periodic climate shift known as El Niño was the oft-cited culprit, a warming climate is making El Niño more severe, more common, and longer-lasting.

Norse never expected marlin and tuna to arrive in the stereotypically cool, rainy region. "That is extremely peculiar," he says.

## A Profound Change

Marlin and melting glaciers are concrete signs that things are changing, but they are not the only harbingers of things to come in a region proud of its symbols—glaciers, orcas, mountains cloaked in fir trees, and wild salmon. As writer Timothy Egan put it, "The Pacific Northwest is simply this: wherever the salmon can get to." Historically, salmon journeyed upstream from the Pacific Ocean into the bowels of the continent to spawn in rivers in Washington, Oregon, northern California, Idaho, Montana, and Canada's British Columbia.

The salmon have been significantly impacted by rising temperatures, as have all of the region's symbols. From the food Northwesterners eat to the power that runs their cities, climate change stands to profoundly affect the region. While no one can definitively tie the current impacts to a human-caused, long-term shift in the climate, many researchers repeat again and again that the changes are consistent with such a warming. Even cautious, middle-of-the-road climate scientist Philip Mote of the University of Washington wants to "underscore" that waiting for proof before taking concrete steps to combat climate change "would not be a prudent course."

Hikers have long enjoyed picnicking in heather meadows at Paradise in Mount Rainier National Park in Washington. But as temperatures rise (the 1990s were warmer than the 1980s, and the 1980s warmer than the 1970s, says Mote), trees are filling in the park's subalpine meadows. The trees are taking advantage of a longer growing season—8 to 10 weeks, compared to 6 to 8, says David Peterson, a USGS researcher and forest ecology professor at the University of Washington.

"Why do people go to Paradise? To see the flowers," Peterson says. "And the flowers are starting to disappear."

Spring is arriving earlier as temperatures rise.

Big clusters of light purple lilacs now begin to bud and release their heady, sweet fragrance on average 8 days earlier than they did back in the 1960s. That is an advance of about 2 days per decade, notes a 2001 study published in in the *Bulletin of the American Meteorological Society*. Honeysuckle vines heavy with the tiny thin-tubed flowers of yellow, red, cream, or purple now begin to bud and give off their sweet, almost cloying scent about 10 days earlier.

Lilac fans noticed as early as 1984 that their favorite flower was blooming earlier than normal in the American West and Northeast. Their discovery sparked interest from scientists such as Daniel R. Cayan, lead author of the American Meteorological Society study and director of the climate research division of the Scripps Institution of Oceanography in La Jolla, California. He soon found that the West was not alone in ushering in an earlier spring. European botanical gardens in the past few decades have noticed flowers are beginning to bloom 6 days earlier and the growing season has extended by 1 to 2 weeks. Quaking aspen trees in the Canadian city of Edmonton, Alberta, now bud 8 days earlier than in the 1930s. So do the shrubby thickets of choke cherry and small deciduous serviceberry.

## Warm-Weather Migrants

Lisa Crozier holds no doubt that warmer weather has led to her distinction of being the first person to discover a particular cold-sensitive butterfly—the sachem skipper (*Atalopedes campestris*)—in the semi-arid desert city of Yakima, located east of the Cascade Mountains in south-central Washington. Hot, dry days and pleasantly cool nights mark the summers of Yakima, population 72,000, and its nearby wine country. While it is blessed with 300 days of sunshine, though, Yakima faces cold winters with average minimum temperatures of 19.7°F.

Crozier says she figured it was just a matter of time before the little cold-sensitive butterfly showed up in Yakima. Typically a southern species, it had already migrated north into eastern Oregon and to a cluster of small communities known as Tri-Cities, Washington, southeast of Yakima, where its caterpillars struggled but hung on through the winter.

She recognized the butterfly instantly when she saw it that day in 1999. "I was very excited," says Crozier, who back then was a University of Washington researcher studying the butterfly. "I would just go every week to see if they were there. I had been doing it for a year. So when they finally appeared, it was thrilling. Initially, I just found one. And then about two weeks later, they appeared more consistently in a whole bunch of different sites."

"They look more like moths," says Crozier, who found a sachem skipper fluttering around the leafy alfalfa and shrubby rabbitbrush she monitored. "So they aren't like the glamorous butterflies that everybody thinks about, but they are butterflies. This particular one is pretty large. They're fairly stocky. They're *extremely* fast fliers—that's why they're called skippers. They fly in a straight line, almost like bees."

Crozier theorizes that the year-round temperatures in Yakima had warmed sufficiently to attract the winged colonizers. She believes a warming climate "was definitely a prerequisite" for the butterfly's arrival. Her studies ruled out several other possible explanations. She is quick to add that "there is *never* just one explanation." Yet, she adds, butterflies are "very sensitive to climate change."

She expects the opportunistic skipper will spread its range further, going over the mountain pass to the west side of the Cascades and into Seattle as temperatures continue to rise faster than they have in 10,000 years—a predicted 2°F by 2020 or 4.5° by 2050, according to climate models at the University of Washington. "They'll be in Seattle eventually," Crozier says. "It just needs to get a little warmer."

## Losing the Snowpack, Losing Water

Warmth is good for the skipper, but it spells trouble for the region's "white gold," as the mountain snowpack has been called, and for anyone dependent on the cool, clear water that rushes down glacier-fed streams in hot July and August. Global climate change threatens to eliminate half the Northwest's snowpack, according to one estimate. Glaciers are "frozen freshwater reservoirs which release water during the drier summer months," Richard S. Williams, Jr., of the U.S. Geological Survey wrote in a report.

Glaciers also are important for tourism. Hartley Brown speaks loudly into his headset so tourists can hear him over the whoopa-whoopa-whoopa sound of his helicopter. He swoops the chopper down into canyons and over a green blanket of coastal rainforest and deftly rises upward to high snowline for the highlight of the 40-minute trip: a view of the Comox glacier on British Columbia's eastern Vancouver Island. More than two hundred times, the pilot for Timberland Helicopters has flown over the popular Canadian glacier. He thinks of it as a giant ice sheet across the mountaintop, though from down below, far far below, people gazing upward tend to think of it as a distant snowcap and picturesque backdrop to the harbor.

Brown sees firsthand that as the glacier shrinks, it exposes rocky patches and new lakes. It also one day could shrink his clientele. As he flies into the high country, he jots notes about the glacier's movements and points out its retreat to his tourists. Before the summer of 2002, he says, "I had never seen our glacier recede back to the point—it's like a little snowcap. Also, one of the lakes up high, which usually only half thaws, completely thawed out."

Not only are glaciers thinning: In the last half-century, the warming climate has made snow "drier"—reducing the water content of the region's springtime snowpack and straining the supply of summertime water for drinking and the irrigation of crops.

Philip Mote, the climate scientist, made that discovery. A wiry guy with a small moustache and even-keel demeanor, Mote calls himself an optimist when it comes to climate change. He rides his bike to work at the University of Washington in Seattle, despite free parking at his building, partly to do what he can to help reduce greenhouse gases, partly to keep trim and sleep well. He attached a trailer to his bicycle to ferry his youngest child to day care until his kids grew old enough to go to school; now he sometimes joins them when they all bicycle 1.25 miles one way to school. Just about every scientist in Mote's department—the Pacific Northwest Climate Impacts Group, a consortium of scientists studying climate change—rides a bike to work or takes public transportation. One colleague does not even own a car. Helping solidify Mote's commitment to commute by bicycle is his latest discovery that the region's snowpack already holds less water. Excitement starts to ripple in his voice.

"I was shocked at how big these declines in snowpack are," says Mote. "There's already a clearer regional signal of warming in the mountains than we expected." He points out that he intentionally cast a wide net when he embarked on his research, so the bad news affects a large audience—people living in Washington, Oregon, Idaho, and Montana west of the Continental Divide, plus the Columbia River basin in Canada's British Columbia. "Just about everywhere, we've seen declines in snowpack of 20, 30 percent. And that's *much* bigger than I expected," Mote says.

The snow that falls onto mountaintops is not just pretty for hikers to look at—it is the summertime water supply for surrounding cities, towns, and rural communities. Less snow in the mountains means less snow will melt throughout the summer to be used as drinking water. Already, parts of the region suffer periodic water shortages. So Mote's discovery presents grim news for cities such as Seattle and Portland. The U.S. Census has predicted that by 2025, another 3.35

million people will live in Washington, Oregon, Idaho, and Montana. That is like adding another state the current size of Oregon, or moving everyone in the cities of Chicago and San Francisco to the Pacific Northwest even as the water supply continues to shrink.

Looking at his figures, Mote notes that from 1950 through 1992, the amount of water contained in snowpack fell steadily throughout the region. All but four of the 145 sites studied had less water in its snowpack, as measured on April 1 of each year. Most declined by at least one-quarter. Nine sites—eight in Oregon and one at Hurricane Ridge on Washington's Olympic Peninsula—had a drop of 60 percent or more. Notably, all of this happened by 1992—before the globe especially heated up. The 1990s became the warmest decade of the millennium, and 1998, 2001, and 2002 were three of the hottest years ever recorded, according to the Pew Center on Global Climate Change. But Mote's snowpack study stops at 1992.

Most of the decline Mote saw traces to higher temperatures that cause the snowpack to melt earlier. At the nine sites with the sharpest decreases, the cause was a combination of rising temperatures and declining precipitation. "The losses generally decrease with elevation, especially in the Cascades, which is consistent with a temperature effect," says Mote, who later presented his findings at the American Meteorological Society's annual meeting in Long Beach, California. "The way that I would portray it is that over the observed period, we have gradually shifted from a fully natural climate regime to one in which humans have a growing influence. That's why we can say that this decline in snowpack points the direction to where we're headed—giving us a glimpse of where we're headed."

But there is more bad news. Not only is snow drier. Not only are glaciers shrinking. An even bigger impact of a warming climate is this: Expect less snow, more rain, with dangerous disruptions of the water supply. Less snow will fall in the first place, and higher temperatures will turn more of that precipitation into rain. Winter rains will melt more accumulated snow earlier in the season, aggravating flooding and affecting drinking water supplies, as well as having a profound economic impact on ski resorts and other snow-dependent businesses.

## Fear of Flooding

It couldn't be happening in a worse place. The Pacific Northwest is already universally known for its rainy climate, and global warming will make winter rainfall heavier and more frequent. As mountain snow is pounded with heavy rain, it melts and unleashes torrents of water down mountainsides, into creeks, into rivers, and increasingly into homes built on floodplains.

Already, Scott Weston sees signs of flooding trouble ahead. Weston, a young geoscientist at Madrone Environmental Services in Abbotsford, British Columbia, points to a series of photos. They show what happened when—to the surprise of neighboring residents—heavy rains caused the Lower Englishman River on eastern Vancouver Island to come alive on March 13, 2003. It rose 2 meters in just 24 hours. "Turns out that this is a trailer park on the flood plain," Weston says, pointing to what looks like a river with buildings in it.

Floodwaters rose so high that water touched the bottom of an outdoor community bulletin board. Weston moves on to the next photo. Here, the water turns a rural subdivision's sole street into a virtual river. It stands high enough to reach a parked car's windshield. So much water roared downriver that day that the Lower Englishman River, which normally discharges 350 cubic feet of water per second, suddenly sent 11,000 cubic feet of water per second careening downriver. To put that into perspective, for several hours enough water roared downstream every second to nearly equal the amount used in an entire year by a typical American household (12,000 to 16,000 cubic feet).

With climate change, Weston predicts more frequent, bigger floods of the Lower Englishman River, just as other scientists predict more floods elsewhere. By 2020, the peak amount of water discharged by the Lower Englishman River could increase by 8 percent a year on average, according to computer models. By 2080, that average would rise to 17 percent.

The likely conclusion: Flooding will reach people's homes more often. Government officials and developers make calculated gambles when they allow development, projecting how often a certain-sized flood will occur based on past weather patterns. The biggest inundation you would expect over a decade is known as a "10-year flood." The whole flood-protection system is calibrated around these averages to make flooding that reaches homes a very rare event. With climate change, though, such flooding is likely to happen roughly twice as often. "What does this mean?" Weston asks. "By the year 2080, a 10-year flood will have the same magnitude as the current 20-year flood event. And a five-year flood will have close to the same magnitude as the current 10-year flood," Weston says.

All of which portends headaches for water managers and people living in floodplains. And the number of such people is likely to grow. "Large areas of the floodplain are currently occupied by houses," Weston says of the Lower Englishman River. "And much of the remaining area is zoned for expansion of housing."

Weston points to a photo of a handsome house sitting squarely on the earth rather than on pylons. "This is a recently built house on the floodplain," he says. "Flooding on a floodplain is a natural phenomenon, and it's only a problem where humans have elected to use these areas."

## Whither the Wild Salmon?

As bad as this less-snow/more-rain trend may be for people, it already poses problems for a beloved regional symbol—wild salmon. "Climate change," according to a report by Canada's David Suzuki Foundation, "is seen as one of the causes of a dramatic drop in Pacific salmon populations along the west coast of North America."

Kim Hyatt speaks matter-of-factly. For more than 25 years, he has made it his life's work to understand the sockeye salmon of Canada's Pacific region, particularly the Okanagan sockeye, the last salmon stock of dozens that formerly returned from the Pacific Ocean to Canada through the Columbia River. About 30 percent of all British Columbia sockeye originate from warmer watersheds near the southern end of the species' range in North America, where these cold-loving, temperature-sensitive fish are highly susceptible to future climate change.

"On the Columbia River, there's a very sad history for sockeye salmon in general. In the late 1800s, there were over half a million fish landed annually," says Hyatt, a scientist with Canada's Department of Fisheries and Oceans, Pacific Biological Station, in Nanaimo, British Columbia. Yet in the last century, "it's declined to very low levels, to the point where there are very few sustainable fisheries."

Salmon already lead a punishing life by human standards, without even considering the effects of climate change. First of all, they essentially quit eating when they are ready to leave the Pacific Ocean to return up rivers to spawn. So they're burning their fat reserves as they journey hundreds of miles up rivers. They swim against the current the whole way, flinging their bodies headlong over waterfalls, boulders, fallen logs, and other obstacles. Chunks of skin, sometimes flesh, hang from their bodies by the time the lucky few salmon that make it this far finally arrive at their ancestral spawning area. Tired, hungry—now it is time to fight. They compete with other salmon for mates. They compete for an ideal particular spot to spawn. Upon mating, they gasp their final breaths and die. With the past century's advent of dams, over-fishing, and the detrimental effects of human development, fish hatcheries, and streamside logging, wild salmon numbers have plummeted. Climate change aside, they face a precarious existence. Hyatt has found that warmer temperatures have made it that much harder for sockeye salmon.

Sockeye live by strict temperature rules. The rules work like this: Adult salmon will stop their arduous journey to their spawning sites whenever the water temperature hits 70°F. It is simply too stressful to continue. They will wait for temperatures to turn. For Okanagan sockeye, anything above 63°F is stressful, and 77°F is lethal. So they wait. And wait. Migrations actually will resume at 72° or 73° so long as the salmon sense that the temperatures have crested and now are definitely declining. It can be a long wait—on average, 40 days, but as long as 72 days. "Delays have been *increasing* steadily in association with climate warming during the 1985 to 2000 interval," Hyatt says. That further tires and stresses the fish before they even begin the final leg of their odyssey. Salmon now spawn about 8 days later due to warming waters. Since 1985, on average, the peak spawn date for Okanagan sockeye has been October 19. During the prior cooler years (1947 to 1985), the average peak spawn date was October 11.

This delay means "you're starting to *push* the ecological envelope that these guys have in terms of available energy resources," Hyatt says. Even for those that make it, this delay bodes ill for their successful reproduction. By spawning later, they leave their eggs more prone to being washed away when the spring snow thaws start and river-scouring floods rip through the gravel where the eggs are laid. Eggs now hatch 15 days later—on average, on March 3. Compare that to February 16, the average peak hatching date back in the cooler years of 1947 to 1985. The longer it takes for eggs to hatch, the likelier it is they will be swept away in springtime floods.

And if that were not enough, juvenile salmon—the next life history stage of the fish—find the going tougher as temperatures warm. They, too, live by strict rules, Hyatt found while studying them at British Columbia's Osoyoos Lake. Rule no. 1: When the fish come up toward the surface to feed at night, they rise only so far. Once they hit the invisible line where the water temperature starts to surpass 63° or 64°F, they stop and will not go beyond it. Rule no. 2: The juvenile fish will swim only in water that contains at least 4 milligrams of oxygen per liter of water. It is "the four-milligram rule," as Hyatt calls it. Colder water retains more oxygen.

"So in the central basin," Hyatt says, the salmon are "squeezed by super-optimal temperatures and sub-optimal oxygens. In the north basin, they still have some latitude to move, but this has profound influences on . . . areas they can use in the lake to feed and grow. The south basin of Osoyoos Lake is unsuitable for rearing through the summer and fall. Sockeye avoid this basin or die."

Warming temperatures have dealt a blow. If the lake's surface warms to 63° early in the season and lingers late, the oxygen progressively depletes. This temperature/oxygen squeeze varies from being severe in years like 2001, to moderate as in 1996, to low as in 1997. Still, on average, the squeeze now occurs more than 90 days per year. Back in the cooler era of 1945 to 1985, the squeeze occurred fewer than 60 days. The lake can become a watery death chamber. In hot 1998, the rearing season began with more than 150,000 juvenile salmon in the central basin of Osoyoos Lake. By mid-October, 95 percent had died.

"They were sealed into the central basin. This temperature/oxygen squeeze occurred, and we measured the mortality at 5.9 percent lost per week—just a straight-line loss," says Hyatt.

So salmon, perhaps the most famous symbol of the Northwest, face trouble every step of the way due to climate change—from migration delays that cause more salmon to die before they can spawn to the reduced viability of their offspring. Salmon need cool water. And it is just plain getting warmer. Compounding matters, rivers around the Pacific Northwest only will warm further as farmers use more water to irrigate thirsty farms and loggers and developers leave too-few streamside trees to help keep streams as cool as possible. There is a wide range of possible ways to try to mitigate problems, involving water storage, water regulations, and other measures, but, Hyatt says, "they're all difficult and most of them are going to be pretty expensive."

Hyatt feels that the story of his Okanagan sockeye has broader implications for dozens or scores of other sockeye stocks around the Pacific Northwest, and the picture is not pretty. Recall that salmon are the primary food for another Northwest icon, orcas. Bear in mind also that salmon returning to Northwest forests bring back the nutrient wealth of the Pacific; fewer salmon carcasses means less-lush forests. And that is on top of British Columbia researcher Dave Spittlehouse's findings that in a warmer, drier world, timber yields of the Northwest's quintessential timber tree, the Douglas fir, will drop along with summer rain levels.

"We're going to be faced with some substantial impacts in the next 25 to 50 years," Hyatt says.

Those effects will reach far past the Great Outdoors and right into Northwesterners' pocketbooks. A big part of this is the double whammy a warmer world will mean for the Northwest's water system. First, as more water runs off in the spring instead of remaining stored in snowpack, operators of the region's dams will have no choice to avoid flooding but to release more water. But once that water flows past a dam, it cannot be used to produce electricity. So, for example, the region's financially troubled power marketer, Bonneville Power Administration, will not have that water available later in the summer, when air conditioners spike demand. Increasingly, Bonneville will have to buy its electricity on the open market. That means higher power bills for households and businesses.

This earlier run-off surge will also hurt farmers. Even now, there are pitched battles over water, sometimes between farmers themselves, sometimes between farmers and fish advocates, occasionally between farmers and cities. The bitter and massive showdown over water in Oregon's Klamath River basin sparked civil disobedience in 2002. Parched farmers, upset that the amount of water they had become accustomed to was reduced because of drought and endangered fish protection, stormed the basin's headgates and turned the water on themselves.

That is unlikely to be the end of the conflict, or the fallout for consumers. Because of earlier snowpack melting, farmers are now facing increased irrigation water shortages in the late summer, which could mean lower crop yields and higher prices for locally grown food.

More trouble is expected. Here are some future scenarios for the Pacific Northwest, according to Patrick Mazza of Olympia-based Climate Solutions, a project of the nonprofit Earth Island Institute: droughts coming twice as frequently by 2020. Forests retreating from the eastern Cascades in Oregon and Washington, replaced by grasslands. Ski seasons shortened. More frequent destructive floods and mudslides. More hot days in summer.

Many scientists are concerned about these looming calamaties. Some are even frustrated and speaking out—expressing thoughts more forcefully than is usual from what is normally a reticent group of experts. Frustration tinges the voice of Edward Miles, a University of Washington marine studies professor and head of the Climate Impacts Group, the consortium of scientists organized by the university and the National Oceanic and Atmospheric Administration.

"The country as a whole is in denial about this problem," Miles says of climate change. "In the meantime, nature's adding up the bill. The world is going to have to deal with this for the next 200 or 300 years. We have never faced this kind of problem before."

Richard Gammon is considered Dr. Doom by some. He recalls putting studded tires on his car in winter, but, thanks to newfound warmer winters in Seattle, he has not had to for several years. "When I see a dying madrona tree in Seattle, I think 'global climate change,'" says Gammon, a University of Washington scientist who believes the local tree is sensitive to the changing climate. "For the last 100 years, the Pacific Northwest has been warming and having increased rainfall. These trends are accelerating now."

It may be tempting to write off Gammon as an alarmist, except for his credentials. He helped author the original United Nations–sponsored Intergovernmental Panel on Climate Change report in 1990 and has reviewed its two succeeding reports. To him, nothing short of the Pacific Northwest's culture is at risk. "If we like cedars, if we like orcas, and bald eagles and salmon—all those things are at risk. When the salmon go, the eagles will go, and the orcas will, too," Gammon cautions. "The Yakima Nation said: 'When the salmon are gone, we're gone.'"

CHAPTER TEN

# Antarctica: The Ice Is Moving

*David Helvarg*

Puking penguins and global warming may not, on first blush, appear to have much in common. But as I discovered in Antarctica, scientific evidence is often where you find it.

Torgersen Island, Antarctica, with its thousands of squawking, flipper-flapping, chick-pecking penguins, combines the pungent odor of a cow-barn with the sound levels of a hip-hop concert. Plus most of the birds here, parents and chicks alike, have managed to stain themselves the color of Georgia red clay with their krill-rich guano.

But down by the water's edge adelie penguins are jumping ashore clean, wet, and plump from the icy Southern Ocean, cute enough to grace the cover of any environmental calendar.

Approaching these full-bellied birds is Dr. Bill Fraser, a rangy, ginger-haired scientist and 25-year ice veteran from Montana State University. Working here on the highest, driest, coldest continent on Earth, he has become one of the world's leading authorities on penguins. He is in his usual uniform of beater bill cap, blue fleece jacket, and deeply stained boat pants and boots.

"Looks like the birds are having problems finding food this year," he tells me. "Normally it takes them six hours, but now we're finding they're spending as much as 16 hours a day foraging for krill."

How does Fraser know the birds' diets and the availability of their prey? One way is a technique called diet sampling.

Fraser approaches one of the foot-and-a-half-tall adelies. Holding its wings out for balance, it veers away to his left. But before it can get more than a few paces, his long-handled net flashes faster than a striking leopard seal, and the bird's hanging upside down 4 feet off the ground.

Fraser reaches into the net and extracts it by a flipperlike wing, walking the bird over to where fellow researchers Matt Irinaga and Donna Patterson have set up their diet-sampling equipment, including a Black & Decker field transfer pump.

Patterson kneels on a padded board, taking the bird between her knees. Fraser has a big insulated jug full of warm saline water around 100°F (the same temperature as the bird's stomach). He dips the tip of an attached plastic tube into mineral oil so as not to hurt the bird's throat and slips it down its gullet. He then runs water through the tube with the hand pump until the bird starts to gurgle. At this point they pull the tube and the bird starts upchucking krill. They hold the bird tail end up until it empties out into a bucket. It does not sound as bad as a drunken freshman at the tail-end of a frat party, more like a pitcher of ice water being poured out. I record the sound on my mini disc recorder for Marketplace Radio, one of the perks of being a far-flung correspondent.

After the crew right the bird, it shakes its head vigorously, getting regurgitated krill on everyone's boat pants and fleece jackets, before it belly slides and paddle walks away, looking somewhat indignant, as penguins often do.

Returning to the bucket, Fraser, Irinaga, and Patterson use kitchen strainers to drain and pack down the post-penguin krill, slipping it into ziplock baggies.

We then head back to our Zodiac, tied up on the north side of the island. Here we collect our orange float coats where we have draped them over several large rocks. I undo the raft's line that is tied around a jagged granite spur. I take a seat on one of the raft's hard rubber pontoons as Fraser fires up the outboard and puts the engine into reverse, crunching through some chunky brash ice that has floated in behind us.

As soon as we are clear of the ice, Fraser puts the raft into forward, spins us in a tight circle, and guns it. "Palmer this is *Schnappers*. We just left Torgersen and are headed back to station," Patterson reports on her hand radio (Fraser has named his boat team after a Wisconsin Polka band).

"*Schnappers*, I roger that," replies the mellow voice of the station's communications man. The air temperature is a mild 34°F, but the wind and spray off the cold, odorless Southern Ocean makes it feel a lot colder, reminding us of where we are, as does a flight of blue-eyed shags racing across the subfreezing chop of the sea.

Squinting through the spray, I can see our destination hunkered down in a boulder field below a blue-white glacier.

Palmer Station is one of three U.S. Antarctic bases run by the National Science Foundation. It is located on Anvers Island, 38 miles of granite rock covered by hard ice up to 2,000 feet thick. Anvers is part of the Antarctic Peninsula, a 700-mile-long tail to the coldest, driest, highest continent on Earth, a landmass bigger than the United States and Mexico combined, containing 70 percent of the world's freshwater and 90 percent of its ice. The peninsula, where polar and marine climates converge, is also a wildlife-rich habitat that researchers refer to as "the banana belt." And that was before global warming.

Palmer itself, where I spent 6 weeks one austral summer, has the look of a low-rent ski resort next to an outdoor equipment dealership.

It is made up of a group of blue and white pre-fab metal buildings, with two big fuel tanks, front loaders, snowmobiles and shipping containers scattered around. The two main buildings, Biolab and GWR (Garage, Warehouse and Recreation), are separated so that if one burns down, the other can act as a refuge for the twenty to forty scientists and support personnel who work here year round.

The short oblong pier on Hero Inlet has giant rubber fenders, where the *Gould*, a 240-foot supply and research vessel, docks every 6 to 8 weeks during the austral summer. January and February's summer temperatures drift between a balmy 0 and 40°F, with 23 hours of daylight to enjoy the views (during winter darkness temperatures average 50 to 100° colder). The weather is always variable with sun, clouds, wind, rain, snow, and gale force winds, often on the same day, kind of like the San Francisco Bay Area on steroids. Next to the pier is the boathouse where we tie up our black 20-foot Mark 5 Zodiac next to a string of other 15- and 20-foot rubber-hulled watercraft. They are ideally suited for getting around in the surrounding ice-studded waters where aluminum-bottomed boats would be crushed like beer cans.

After offloading our boat and gear, drying out, and grabbing some food, Fraser's crew heads to Biolab's work area adjacent to the aquarium tanks.

"Some of our birds are eating *Chysanoessa Macrura*, a smaller species of krill, which means they might be having a hard time finding their regular prey," Fraser explains to me when I catch up with them. He is using a tweezers to point into a tray full of partly digested krill on the lab table in front of him. Patterson and Irinaga are tweezering through similar trays.

They need to analyze about fifty little pink shrimplike krill to get a good representation of each bird's diet. The krill Fraser has collected looks blacker and lumpier than Irinaga's or Patterson's because they are from different species and more digested, he explains. I decide to skip the jambalaya listed as tonight's main course on the cafeteria blackboard.

For the past 35 years, climatologists have predicted that global warming linked to the burning of CO2-rich fossil fuels would occur most rapidly at the poles, a fact now confirmed by scientists in Alaska,

Canada, Greenland, at the North Pole, and here on the Antarctic Peninsula.

One way that we know there is more heat-trapping $CO_2$ in our atmosphere today than at any time in at least the past 420,000 years is through ice-core samples taken from Siple Dome, Vostok, and other sites in the Antarctic interior. These ice cores contain trapped bubbles of ancient air that have been isolated, dated, and chemically analyzed. They also show that climate is far less stable than we have imagined, and that the past 10,000 years—the period that has seen the rise of human civilization—has also been a period of atypical climate stability. "Climate's a dangerous beast and we're poking it with a sharp stick," is how one scientist described our contribution of industrial carbon to the atmosphere.

Rapid warming in the Antarctic Peninsula over the last 50 years, including an incredible 10°F rise during the austral winter months, has led to a decline of winter sea ice that krill depend on for their productivity. The underside of the ice acts like an upside-down coral reef, providing young krill both food and protective shelter. But heavy winter sea ice that used to appear 4 out of 5 years declined in the 1990s to only 1 or 2 years out of 5. If the krill population declines along with the sea ice that could wipe out populations not only of penguins, but also seals and whales that depend on the krill for their survival. An average blue whale consumes more than 4 tons of krill per day.

The Antarctic Peninsula has been making other kinds of climate news of late as huge pieces of the Larson-B ice shelf—including one berg twice the size of the state of Delaware—have begun calving off its eastern shore.

Scientists are now discussing the possibility that the western Antarctic ice sheet adjacent to the peninsula could experience a sudden meltdown, raising global sea levels by 18 to 20 feet (instead of the 1 to 3 feet currently predicted by 2100). While most experts believe this melting will occur sometime after the twenty-first century, by the time they know for sure, it will be too late to do anything about it.

Other dramatic signs of climate change include retreating glaciers and increased snowfall (warming in Antarctica means more precipitation in the form of snow). There is also the displacement of ice-dependent species such as adelie penguins, crabeater seals (that actually eat krill), and leopard seals (that eat krill, penguins, other seals, and the occasional Zodiac bumper) by more adaptable open-water species such as chinstrap penguins, elephant seals, and fur seals.

After the Schnappers have finished sorting their krill, I take a walk with Fraser, admiring the ice cliffs of Arthur Harbor and the massive glacier that curves around and above Palmer Station.

"The Marr glacier used to come within 100 yards of the station," Fraser tells me, pointing upslope. "Its melt water was the source of our fresh water." Today the Marr is a quarter-mile hike from Palmer across granite rocks and boulders. Gull-like brown skua birds now splash in the old melt pond, while the station is forced to use a saltwater intake pipe and reverse osmosis desalinization to generate its own freshwater. Periodically the artillery rumble and boom of moving ice alert us to continued glacial retreat and allow for spectacular views of irregular ice faces collapsing into the adjacent harbor, setting off a blue pall of ice crystals and a rolling turquoise wave beneath a newborn scree of chunky brash ice.

On another day I hike with Fraser past dozens of lazing elephant seals, before climbing hundreds of feet up the jagged granite and basalt boulders of Norsel Point, where we are dive-bombed by skuas. We reach a spectacular mossy overview above Loudwater Cove across an open channel from the blue ice of the Marr Glacier. Fraser points to three rugged granite islands that have emerged from below the retreating glacier in recent years, while we listen to the distant artillery rumble of new calving ice.

"When I was a graduate student we were told climate change occurs, but you'll never see the effects in your lifetime. But in the last 25 years I've seen tremendous changes. I've seen islands like these pop out from under glaciers, I've seen species changing places and landscape ecology altered."

Two weeks later there is a surprise awards ceremony at the regular Wednesday night science talk in Palmer Station's rec room. The U.S. Board of Geographic Names, with a nudge from the National Science Foundation, has approved the naming of the largest of the three glacier exposed islands Fraser Island.

Fraser is amused and, also one suspects, quite moved as he is handed a framed illustrated certificate and digital photo of Fraser Island. "Now climate change will reverse and in 25 years it'll be covered up again," he jokes to appreciative laughter.

But everyone in the room knows better.

The science on anthropogenic (human-enhanced) climate change has become well established by now, despite efforts by the fossil-fuel industry (the largest industrial combine in human history) first to deny the science and later to delay the transition from coal and oil to non-carbon-based renewable energy systems, arguing this would be "too costly."

In early 2003, the science journal *Nature* published two studies that found global warming is forcing species around the world to move into new ranges or alter habits in ways that could disrupt their ecosystems and endanger their survival.

In some cases, species' ranges were found to have shifted 60 miles or more in recent decades, mainly toward the poles. In other species the timing of egg laying, migrations, and the like has shifted weeks earlier in the year (with earlier springs), creating the risk that species could be separated in both time and space from their needed sources of food.

Fraser has found that by being hardwired not to shift their egg-laying patterns or locations, adelie penguins may face a similar risk of climate-driven extinction.

We are on Humble Island, a 20-minute boat ride from Palmer, walking toward a wide flat field of pebbles that is actually a graveyard. We walk past a dozen burbling 800- to 1,000-pound elephant seals laying in their own green waste. One of them rises up just enough to show us a wide pink mouth and issue a belching challenge: Stay back or I may have to rouse myself from complete stupor in order to attack you. The elephant seal population, once restricted to more northerly climes, is now booming along the peninsula, while adelie populations are crashing.

"These penguins are the ultimate canaries in the mine shaft. They're extremely sensitive indicators of climate change," Fraser tells me, explaining how the open pebbled area we are standing on represents an abandoned adelie colony (they use pebbles as nest-building material). He has been studying seventy-two colonies in a 40-square-mile area for 25 years now and has watched at least one colony a year go extinct since 1985. A population of 15,200 breeding pair of adelies has decreased to 9,200 pairs, while what were only six pairs of chinstrap penguins has increased to 360 pair.

"So here you have two identical krill predators," Fraser explains. "The question we asked ourselves was why the different trends?"

It is a mystery he thinks he has solved.

"On a large scale, we see winter sea ice declining, but on the local scale we also see warming creating more precipitation as snow," he says. "With storms coming out of the north and northeast, we see this snow accumulating on the southern, leeward side of islands, which is where the adelie colonies are all going extinct. These birds need dry ground to lay and hatch their eggs. But the increased snow's altering the available nesting and chick-rearing habitat."

We walk through another dying colony with only a few birds and chicks where once there were hundreds. Overhead, skuas are gliding below a cloudy sky, looking for weak or isolated chicks to feed on.

"Once their numbers decline to this point the scavengers have a field day," Fraser notes. "Chinstraps breed later in the season after the snow's melted and so do better. They're a weedy species. They adapt well to disturbed habitat and can also take fish and squid when krill aren't available because they can dive deeper and feed at night. They're the dandelions of the penguin world."

In a warmer world, "weedlike" species that are highly adaptable to disrupted habitat (pigeons, rats, raccoons, deer, elephant seals, and chinstrap penguins) will displace more specialized "endemic" creatures (tigers, monarch butterflies, river dolphins, sea turtles, and adelie penguins) that depend on unique ecosystems such as rain forests, coral reefs, and the Antarctic ice shelf.

In this way climate change is speeding up a global chain reaction of extinctions that—thanks to other impacts from humans—is already under way and has been labeled the sixth great "extinction pulse" in planetary history. The last one, a meteor-based event, took out the dinosaurs some 60 million years ago.

Half an hour later, I am waving off Antarctica's answer to a pterodactyl, a dive-bombing skua. Somewhere along the line these birds failed to get the word that they are not at the top of the food chain. They will go after you if you get too close to their nests, startle them, or just look at them the wrong way.

Meanwhile, Fraser is conferring with Rick Sanchez of the U.S. Geological Survey. Sanchez is carrying a portable Global Positioning System (GPS), along with a satellite antenna sticking out of his backpack and a magnesium-shelled laptop strapped to an elaborate fold-down rig hanging from his waist and shoulders. He is trying to walk off the perimeter (he calls it the polygon) of an extinct colony of adelies in order to confirm Fraser's observations, but a burbling pile of elephant seals is blocking his mapping venture. If he tries to move them, they might stampede and crush still-living penguin chicks or perhaps a curious scientist. Such are the quandaries of high-tech research projects in Antarctica.

## Petrels in Peril

I climb up to the island's higher crags and ridges to join Donna Patterson who is checking on her favorite birds. Giant petrels are large scavengers, about the size of bald eagle, that feed on dead seals, whales, penguins, squid, or just about any other creature that dies on Antarctica's Southern Ocean. This is why nineteenth-century sailors called them "bone shakers." They also called them "stinkers" because, if challenged, they can spit a vile fishy stomach oil up to 6 feet.

Patterson has another view of giant petrels or "GPs" (pronounced "Jeeps"). "I think giant petrels rule the earth, they're fantastic," she tells me. Patterson is 5 feet 1 with a bright but determined attitude and sun-reddened face (except for a raccoon-like band of white around her eyes from wearing sunglasses and a bill cap in the field). She has spent the last 7 years in the company of giant petrels.

We approach the first of twenty-eight birds she will examine today. It is sitting on a nest of stones, weighing in at about 12 pounds with a long tube bill hooking to a sharp down turned point tough enough to puncture seal skin. Patterson crouches down by the bird, talking quietly as the gray-and-white feathered creature begins cawing and clacking at her, spreading its 6-foot wing span as she reaches underneath it and pulls out a downy white chick.

The brooding bird settles back down as Patterson measures the length of the chick's culmin (upper beak), which is the best indicator of its age, and then slips it into a net bag for weighing.

As she is placing the chick back underneath the bird, it turns its head sideways to peek under Patterson's bill cap. "Some think that my billcap is my beak. Others know that I'm not like them. They'll look under the bill and test my ears and nose with their culmin," she says. "I always wear sunglasses because they can get a little nippy."

The next bird she visits quietly vocalizes to her the way it would to another bird. As Patterson is returning this one's chick, it takes her sleeve and weigh bag in its bill and begins pulling on them. "This one's so broody she wants to take me and the bag under her," she explains. "They've taken my gloves and mittens and placed them underneath themselves."

Of course, giant petrels are rarely this calm in the presence of humans. Ordinarily they are more likely to spit than coo. It took Patterson years of approaching their nests every day before the birds would let her handle their chicks.

"Do you name the birds?" I wonder. "No, no . . . well yeah, Norman Bates, I've named him. He's just an angry, psycho male."

Both male and female GPs share in the incubation, brooding, and feeding of their chicks. They lay only one egg at a time, and if that egg is damaged, they will not relay in the same season. Within 16 days of hatching both parents fly off in search of food, while the growing chick waits alone for them at the nest. By this age a healthy young "stinker" can defend itself by projectile vomiting its fishy stomach oil at skuas and other potential predators.

Aside from being nipped, Patterson faces other challenges in her fieldwork, including dive-bombing skuas, rough seas, sudden snowstorms, and charging fur seals. "I've been scrambling up rocks and stuck my head over a rise and found myself face to face with a fur seal. They've gone after me," she recalls with a nervous grin. Still her work is helping to expand the world's knowledge of these rare creatures.

"I believe these are valuable birds that can tell us a lot," Patterson explains. "They're individually long lived. I'm working with a breeding female that was first banded in 1965. They're also small in number, about 50,000 breeding pair in total." She points out some penguin bones near one of the bird's nests. "I found a penguin heart there yesterday. Around the nests we've also found seal skin, squid beaks, krill."

I shake my head sadly, poking at the penguin bones.

"They're doing their job cleaning up the Southern Ocean," Patterson says somewhat defensively.

Along with rocks, many of the giant petrels make their nests of shell. Both the rock and shell nests have a surprisingly sculptural appearance, pleasing in their symmetry.

Patterson tracks each of her birds by island of origin, nest number, and band number (for leg-banded birds). One of the things these censuses have helped establish is the recent decline of the species' population. As many as 100,000 Antarctic seabirds, mainly albatrosses, but also giant petrels, are thought to be dying every year in encounters with long-line fishing fleets that have expanded their operations into the Southern Ocean. The birds dive for the baits as the long lines, often containing several thousand hooks, are unreeled from the high-seas fishing boats. If the birds catch the bait at the surface, they are often dragged under and drowned, or else entangled and injured.

"We've found hooks and lines around the nests as well as hooks engorged in the bird's throats," Patterson reports. As a result, slow-to-breed giant petrel populations are now crashing along with various species of albatross.

Much of the long-line fishery threatening the birds involves the unregulated and often illegal taking of Patagonian toothfish, a deep-water Antarctic species that biologists say is also being pushed toward extinction by over-fishing. Toothfish are frequently mislabeled, shipped, and sold in the United States as Chilean Sea Bass.

Another concern in Patterson 's study is climate change. "With this warming trend we're seeing more elephant seals moving south and an increase in elephant seal populations could prove disastrous for nesting petrels because of the competition for space," she explains. "Also warming has increased snowfall, and if there's too much snow that could reduce their nesting sites."

"These birds are good indicators of the health of this ecosystem," Patterson adds, weighing another white fluff ball of a chick. "But if we let them go extinct because we don't value them, what does that tell us about our own species?"

## Harm to Hairgrass

It is not just Antarctica's bird and animal species that are seeing climate-driven shifts in population and dominance, however.

Tad Day is a sandy-haired, boyish, 39-year-old professor from Arizona State University, who drives his Mark 3 Zodiac as if it was a Formula One race car. He has also spent years studying Antarctica's only two flowering plants, hairgrass and pearlwort. The main study site he and his "Sundevils" use is Stepping Stone Island, a surprisingly green, rocky isle several miles south of Palmer around the rough chop of Bonaparte Point.

"Step" is surrounded by pale blue icebergs, a rumbling blue-white glacier, and other rocky islands and outcroppings—including Biscoe Point to the south, which, with the retreat of the Marr Glacier, has now become Biscoe Island. Amidst nesting giant petrels and a friendly skua named "Yogi," Day maintains two gardens, fenced to keep fur seals out, containing over ninety wire plant frames surrounding banks of hairgrass and pearlwort growing not in true soil, but a close approximation made up of glacial sand and guano.

Here Day has found that warming improves the growth of pearlwort but appears to have a negative impact on hairgrass. Once the dominant species in Antarctica, hairgrass is now being displaced by pearlwort, a mosslike plant.

"Global warming," Day explains, "has the capacity to shift the competitive balance of species in ways that, until we get out there and do the research, we don't understand yet, and that could have important consequences on our ability to produce food and fiber."

Increasingly reliable climate models now predict a 3 to 9° planetary warming this century. This will result in shifts in agricultural production, spread of tropical insects and diseases, increases in extreme weather events, more intense coastal storms and hurricanes, erosion of beaches, coral bleaching, and rising sea levels.

So I could argue that I am learning a survival skill when I learn how to operate the Zodiacs, which, along with thick Sorel boots and ice crampons, provide the main means of transport at Palmer.

Taking a 15-foot rubber boat through floating fragments of brash ice, I spot a leopard seal lazing on an ice floe. I maneuver around to take some photos of the snaky, blunt-headed predator. As I am doing this, a panicked penguin jumps into my boat, tripping over the outboard's gas can in front of me. We exchange looks of mutual bewilderment before it leaps onto a pontoon and dives back into the icy blue water.

Incidents like this make it hard to maintain a sense of gloom and doom on the last wild continent, at least for more than a few hours at a time. Along with nightly discussions over Pisco Sours made with glacier ice at the Penguin Pub (the open bar located above the machine shop in GWR), I also manage to distract myself with recreational sojourns on the Southern Ocean. You need at least two people with radios to take out one of the Zodiacs, so on days when I'm not working with the Schnappers or the Sundevils, I spend time looking for a boating partner. Doc Labarre, the station's big, balding fatalistic physician who used to work the emergency room in Kodiak, Alaska, before it was taken over by an HMO, is among those regularly up for an adventure.

One day we cruise past Torgersen Island, where I take the Zodiac up "on platform" (as you speed up, the bow drops down, giving you greater visibility and control), and head us toward Loudwater Cove on the other side of Norsel Point. The following seas allow us to surf the 15-foot craft past the rocky spires of Litchfield Island and around the big breaking waves at Norsel. We then motor around a few sculptural apartment-sized icebergs, crossing over to a landing opposite the glacier wall. We tie off our bowline, watching a big lanky leopard seal sleeping on an adjacent ice floe. We dump our orange float coats and climb several hundred feet up and over some rocky scree and down a snowfield splotched with red algae to the opening of an ice cave. Down in the cave, it is like a dripping blue tunnel, with slush over a clear ice floor that reveals the rocky piedmont. Hard blue glacier ice forms the bumpy roof, with its stalactitelike icicles and delicate ice rills forming pressure joints along its edges. Outside, we hike the loose granite, feldspar, and glacial sand until we encounter a fur seal hauled up, several hundred yards from the water on the sharp broken rocks. He barks and whines a warning at us. Nearby ponds and 100-year-old moss beds have attracted colonies of brown skuas who soon begin dive-bombing us. I get whacked from behind by one of the 5-pound thugs. It feels like getting slapped hard in the back of the head by a large man. We quickly move away, climbing back over the exposed glacier rock, past middens of limpet shells, and down a rock chimney to where our boat is tied up. The leopard seal is awake now, checking us out as we take off. I notice bloodstains on the ice where he has been resting.

We next drop by Christine, a big bouldery island where we walk past a large congregation of elephant seals hanging out opposite a colony of squawking adelies. Crossing the heights, we find mossy green swales with ponds full of brine shrimp. We then lay out on a rocky beach at the end of a narrow blue channel, sharing the space with two elephant seals of about 500 and 900 pounds.

The Southern Ocean is crystal clear; the sun has come out and turned the sky cobalt blue. It feels almost tropical, lounging to the sound of the waves rolling and retreating across smooth fist-sized stones. Further out are several flat islands with big breakers crashing over them, sending spray 50 feet into the air. The elephant seals are blowing snot and blinking their huge red eyes, their black pupils the size of teaspoons for gathering light in deep-diving forays after squid. A fur seal comes corkscrewing through the channel's water before paddle-walking ashore and scratching itself with a hind flipper, a blissful expression on his wolfy face. And there we are, just five lazy mammals enjoying a bit of sun.

Driving the Zodiac back to station, we are accompanied by a flight of blue-eyed shags and squads of porpoiseing penguins in the water. Doc steers while I keep an eye out for whales, like the minke that bumped the boat I was riding in a few days earlier. (It was a real Melville moment, watching its huge brown back roll out from under our raft.) The sky has again quilted over with clouds, turning the water the color of hammered tin; and with the buck and slap of the boat and the icy cold saltwater spray in our faces, it feels as if all is right with the wild.

Antarctica is vast and awesome in its indifference to the human condition. At the same time it has become a world center for scientific research and provided us fair warning about the human impact of global warming on our climate. The message from the ice is as plain as the bones I found scattered around a dying penguin colony. Our world, and theirs, Antarctica's Southern Ocean and America's Blue Frontier, the Antarctic Peninsula and the Jersey Shore, are more closely linked than we imagine.

# Author Biographies

**Sherry Barnes** currently lives on Galankin Island near Sitka, Alaska. A graduate of Prescott College in Arizona and a former intern for *E/The Environmental Magazine,* she has produced radio series for National Public Radio in Alaska, and is currently working at a faculty support center for the University of Alaska Southeast.

**Gary Braasch** is a nature photojournalist who covers environmental issues for magazines worldwide. He was named "Outstanding Nature Photographer" in 2003 by the North American Nature Photography Association. He is currently documenting the effects of climate change on an international scale for his project, "World View of Global Warming," which maintains a popular archive of photos and scientific information at www.worldviewofglobalwarming.org.

**Sally Deneen** is a prolific Seattle-based freelance writer whose work has appeared in *E/The Environmental Magazine, Columbia Journalism Review, U.S.News & World Report, Boca Raton,* and numerous other magazines. The Florida Magazine Association named her "Writer of the Year" in 1998.

**Ross Gelbspan** was an editor and reporter for 31 years at the *Philadelphia Bulletin*, the *Washington Post*, and the *Boston Globe*. At the *Globe*, he conceived, directed, and edited a series of articles that won a Pulitzer Prize. In 1998, he published *The Heat Is On: The Climate Crisis, the Cover-up, the Prescription* (Perseus Books). His extensive speaking engagements have included presentations at, among other places, the World Economic Forum, Renaissance Weekend, a Conference of Socialist Scholars, a climate summit of sixty colleges and universities, and a recent conference of foundation executives. Gelbspan has appeared on *Nightline*, *All Things Considered*, and *Talk of the Nation*, among other radio and television programs. He has briefed the executives of oil companies in the United States, Egypt, and elsewhere. Gelbspan's articles have appeared in *E/The Environmental Magazine*, *Harper's*, *The Atlantic Monthly*, *The American Prospect*, *The Nation*, *Sierra*, *Grist* and a number of other newspapers and magazines in the United States and abroad. He maintains a website at http://www.heatisonline.org.

**David Helvarg** is president of the Blue Frontier Campaign (bluefront.org) and the author of two books, *Blue Frontier: Saving America's Living Seas* (Owl Books, a *Los Angeles Times* "Best Book," 2001) and *The War against the Greens* (Sierra Club Books, 1994). He has worked as a war correspondent in Northern Ireland and Central America, and covered a range of issues, reporting from every continent including Antarctica. An award-winning journalist, Helvarg has produced more than forty broadcast documentaries for PBS, The Discovery Channel, and other networks. His print work has appeared in publications including the *New York Times*, *Los Angeles Times*, *Smithsonian*, *Popular Science*, *Science*, *Sierra*, and *The Nation*. He has led training workshops for fellow journalists in Poland, Turkey, Tunisia, Slovakia, and Washington, D.C. He is a regular commentator for Marketplace Radio and a licensed private investigator.

**Mark Hertsgaard** is a journalist with National Public Radio and Worldlink TV. He is the author, most recently, of *The Eagle's Shadow: Why America Fascinates and Infuriates the World* (Farrar, Straus & Giroux, 2002) and *Earth Odyssey: Around the World in Search of Our Environmental Future* (Broadway Books, 2000).

**Orna Izakson** is a freelance environmental journalist specializing in natural resources, endangered species, ecology, and science. Her award-winning coverage of climate-change issues has appeared in newspapers on both U.S. coasts as well as in national magazines including *E/The Environmental Magazine*. She is currently at work on a book on the crisis in the Klamath Basin of southern Oregon and northern California. She makes her home along the Columbia River in Portland, Oregon.

**Jim Motavalli** is editor of *E/The Environmental Magazine*, and author of two previous books, *Forward Drive: The Race to Build "Clean" Cars for the Future* and *Breaking Gridlock: Moving Toward Transportation That Works* (both Sierra Club Books, 2000 and 2001). He is a regular contributor to the *New York Times*, Cleveland's *Plain Dealer*, *The Nation*, and many other periodicals, and writes a weekly syndicated auto column, as well as regular columns for Environmental Defense and *AMC Outdoors*. He is also the host of a radio program on noncommercial WPKN-FM in Connecticut.

**Kieran Mulvaney** lives in Anchorage, Alaska, and is the author of *At the Ends of the Earth: A History of the Polar Regions* (Island Press, 2001) and *The Whaling Season: An Inside Account of the Struggle to Stop Commercial Whaling* (Island Press, 2003).

**Dick Russell** is the author of three books, including *Eye of the Whale* (Simon and Schuster), named a Best Book of 2001 by three major newspapers. He divides his time between Boston and Los Angeles.

**Colin Woodard** is a self-employed writer and journalist and author of *Ocean's End: Travels through Endangered Seas* (Basic Books, 2001), a narrative non-fiction account of the deterioration of the world's oceans. He has reported from more than thirty foreign countries and six continents for the *San Francisco Chronicle*, the *Christian Science Monitor*, the *Chronicle of Higher Education*, and many other publications. His second book, *The Lobster Coast: Rebels, Rusticators, and the Struggle for a Forgotten Frontier*, a cultural and environmental history of coastal Maine, will be published by Viking in 2004. He lives in Portland, Maine, and maintains a website at http://www.colinwoodard.com.

# Endnotes

## PART ONE: HUMAN IMPACTS

### Chapter One

The firsthand observations and direct quotations in this chapter are by and large based on the author's 6-week visit to China in the winter of 1996 to 1997. For a more detailed account of this journey, and the related environmental questions examined, as well as the documentary sources for other factual points made in this chapter, please see the author's book *Earth Odyssey: Around the World in Search of Our Environmental Future* (New York: Broadway Books, 1999). Additional research and reporting since then drew on the invaluable website maintained by the China Energy Project of the Lawrence Berkeley National Laboratory of the University of California, which contains links to the equally useful Beijing Energy-Efficiency Center. Sources for most other facts contained in this article should be evident from the text. For any additional questions, please feel free to contact the author via his website http://www.markhertsgaard.com.

### Chapter Two

Colin Woodard gathered material for this chapter while on reporting trips to Europe in 2001 and 2002. Much of his original reporting can be accessed electronically through links at http://www.colinwoodard.com/articles. He thanks the Journalism Fellowship Program of the German Marshall Fund of the United States for its support.

The respective Dutch and Venetian flood control plans are presented in the following documents, all of which are available in English: "Rising Waters: Impacts of the Greenhouse Effect for the Netherlands," The Hague: Rijskwaterstaat, January 1991; "Twice a River: Rhine and Meuse in the Netherlands," The Hague: RIZA, 1999; "Measures for the Protection of Venice and Its Lagoon," Venice: Consorzio Venezia Nuova, August 1997; Atlas of the Works Quarterly, Year VIII, No. 1/2, Venice: Venice Water Authority, January 2000. The central arguments against the Venice plan appear in Albert J. Ammerman and Charles E. McClennen, "Saving Venice," *Science* 25, August 2000, p. 1301.

The history of Venice's fight with the sea may be found in Jonathan Kehaey's *Venice against the Sea* (New York: St. Martin's Press, 2002). Visitors to the Netherlands can learn a great deal about the history of the Dutch battle with the sea at the NieuwLand Poldermuseum in Lelystad. Further information on Europe's wind industry can be accessed through the European Wind Energy Association at http://www.ewea.org.

## Chapter Three

Jim Motavalli would like to thank the scientists Dr. Paul Epstein of Harvard University, Dr. Janine Bloomfield of Environmental Defense, Vivien Gornitz and Cynthia Rosenzweig of NASA's Goddard Institute for Space Studies, Dr. Dickson Despommier of Columbia University, and Orrin Pilkey of Duke University. Also invaluable in preparing the chapter were Dery Bennett of the American Littoral Society (on the web at http://www.alsnyc.org); Brian Unger of the Surfers' Environmental Alliance (http://www.damoon.net/sea); Scott L. Douglass of the University of South Alabama (author of the useful *Saving America's Beaches*); Andy Wilner and attorney Deborah A. Mans of the NY/NJ Baykeeper (732-291-0176, http://www.nynjbaykeeper.org); the many-hatted Noreen Bodman, president of the business-oriented Jersey Shore Partnership; and kayaker supreme Randall Henriksen. Information on kayaking around Manhattan is available from Henriksen's New York Kayak at 212-924-1327 or online at http://www.nykayak.com.

Useful books on global warming and the coast include Cornelia Dean's *Against the Tide: The Battle for America's Beaches* (New York: Columbia University Press, 1999); Orrin H. Pilkey and Katharine L. Dixon's *The Corps and the Shore* (Washington, D.C.: Island Press, 1996); Pilkey and Wallace Kaufman's *The Beaches Are Moving: The Drowning of America's Shoreline* (Durham, N.C.: Duke University Press, 1979); Lynne T. Edgerton's *The Rising Tide: Global Warming and World Sea Levels* (Washington, D.C.: Natural Resources Defense Council/Island Press, 1991); and *Saving America's Beaches: The Causes of and Solutions to Beach Erosion* by Scott L. Douglass (River Edge, N.J.: World Scientific Publishing, 2002).

A draft of the report "A Wetlands Climate Change Impact Assessment for the Metropolitan East Coast Region" by Ellen Kracauer Hartig, Frederick Mushacke, David Fallon, and Alexander Kolker and prepared for the Center for Climate Systems Research at Columbia University is available online in PDF format at http://metroeast_climate.ciesin.columbia.edu/reports/wetlands.pdf.

The homepage for the 2001 "Climate Change and a Global City: An Assessment of the Metropolitan East Coast Region" report is at Columbia University, http://metroeast_climate.ciesin.columbia.edu. The executive summary and full synthesis report can be downloaded from there.

The Environmental Protection Agency's James G. Titus is author of the study "Greenhouse Effect, Sea-Level Rise and Barrier Islands: Case Study of Long Beach Island, New Jersey," available online at http://users.erols.com/jtitus/NJ/CM.html.

The Environmental Defense report "Global Warming: Sea-Level Rise and the New York Metropolitan Region" by staff scientist Janine Bloomfield (with Molly Smith and Nicholas Thompson) can be downloaded in PDF form at http://www.environmentaldefense.org/pdf.cfm?ContentID=493&FileName=HotNY.pdf.

Harvard University's Center for Health and the Global Environment may be contacted at 617-384-8530, or online at http://www.med.harvard.edu/chge. The interested reader can reach Clean Ocean Action, which focuses on ocean dumping, in Sandy Hook at 732-872-0111, or online at http://www.cleanoceanaction.org.

## Chapter Four

Dick Russell gathered most of the material for this chapter in extensive interviews conducted on a trip to the islands of Antigua and Barbuda in January 2003.

Useful background material came from *Antigua and Barbuda's Initial National Communications on Climate Change*, prepared under UNDP Project ANT/97/G31/1G/99, May 2001; *Antigua and Barbuda's First National Report to the Convention on Biological Diversity*, prepared under UNDP Project ANT/97/G31/1G, Biodiversity Enabling Activity Project; *Environmental Agenda for the 1990's, a Synthesis of the Eastern Caribbean Country*

*Environmental Profile Series*, a publication of the Caribbean Conservation Association and the Island Resources Foundation, September 1991; *Antigua, Barbuda, Redonda: A Historical Sketch* by D. V. Nicholson, Museum of Antigua and Barbuda, and material provided by Antigua's Environmental Awareness Group and Meteorological Office. Also useful were the executive summary of the "Small Island Countries Dialogue on Water and Climate" presented at the World Water Forum, Kyoto, Japan, March 3–16, 2003, and the Environmental News Service article "Vulnerable Caribbean Nations Prepare for Global Warming," June 4, 2001.

The author would especially like to thank Ambassador Lionel Hurst, Carole McCauley of the Environmental Awareness Group, and Tony Johnson of the Siboney Beach Resort for their assistance in providing key contacts on the islands.

### Chapter Five

Jim Motavalli journeyed to India in 1999 to accept *E/The Environmental Magazine*'s "Global Media Award" for population reporting from the Population Institute. The trip included visits to New Dehli, Agra, Bombay, and many of the other cities under the Asian cloud.

The BBC, *Wall Street Journal*, the Indian magazine *The Week*, CNN, and *The Guardian* have all done excellent work on the Asian cloud. The *Journal*'s story "Soot Storm: A Dirty Discovery over Indian Ocean Sets Off a Fight" by John J. Fialka (May 6, 2003, p. A1) discusses the political ramifications of Professor Veerabhadran Ramanathan's work.

The report "The Asian Brown Cloud: Climate and Other Environmental Impacts," prepared by the Center for Clouds, Chemistry and Climate at the University of California, San Diego, is available from the United Nations Environmental Programme at http://www.rrcap.unep.org, or by calling 662-516-2124.

Columbia University's Center for International Earth Science Information Network (CIESIN) hosted the "Photo-Oxidants, Fine Particles, and Haze across the Arctic and North Atlantic: Transport Observation and Models" conference. Workshop presentations may be accessed at http://www.ciesin.org/pph/agenda.html.

A huge resource for material on Asian haze and smoke is http://www.vadscorner.com/haze.html (though not all the links work!). One that does, at http://www.vadscorner.com/mask.html, illustrates the proper way to strap on a respirator. The collaborative report on the Asian dust events of 1998 is online at http://capita.wustl.edu/Asia-FarEast.

John Hayes's moving account "Life under the Asian Brown Cloud" is posted at http://www.guardian.co.uk/globalwarming/story/0,7369,781095,00.html.

Also well worth reading is the detailed reporting on the human face of climate change by the India Resource Center (formerly CorpWatch India), available on the website http://www.indiaresource.org. The interested reader may access a report from the Climate Justice Summit at http://www.corpwatchindia.org/issues/PID.jsp?articleid=3043.

### Photo Essay

The photographs presented here are part of *World View of Global Warming*, Gary Braasch's documentation of the effects and the science. He interviewed each of the scientists represented in his photo section. More photographs and annotated references may be found at the project website, http://www.worldviewofglobalwarming.org.

There are more than 3,000 individual studies of ongoing effects of global warming on earth systems and the biosphere; 2,500 were reviewed by the Intergovernmental Panel on Climate Change (IPCC), whose report, "Climate Change 2001: Impacts, Adaptation, and Vulnerability," is at http://www.ipcc.ch.

Other important reviews of scientific research with copious references are "Climate Extremes: Observations, Modeling, and Impacts," by David Easterling, et al, in *Science* 289, September 22, 2000; "A globally coherent fingerprint of climate change impacts across natural systems" by Camille Parmesan and Gary Yohe in *Nature* 421, January 2, 2003; and "Ecological responses to recent climate change," by Gian-Reta Walther, et al, in *Nature* 416, March 28, 2002.

For polar regions, please see "Trouble in Polar Paradise" (various authors), *Science* 297, August 30, 2002. For worldwide mountain glacier retreat and sea-level rise, the best sources are "Mass balance of mountain and sub-polar glaciers" by Mark Meier and M.B. Dyurgerov in *Arctic and Alpine Research* 29 (4), 1997, and a second paper by the two in *Science* 297, July 19, 2002. Specific to Antarctic Peninsula changes is "Marine Ecosystem Sensitivity to Climate Change" by Raymond Smith, et al, in *BioScience* 49(5), 1999. And for Alaska, see "Observational Evidence of Recent Change in the Northern High-Latitude Environment" by Mark Serreze, et al, in *Climatic Change* 4, 2000.

A good source on rising temperatures, health and disease is "U.S. National Assessment of the Potential Consequences of Climate Variability and Change: Human Health," available at http://www.usgcrp.gov/usgcrp/nacc/health/default.htm, especially the brochure for downloading, suitable for non-scientists. For information about the European heat wave, please see http://www.earthpolicy.org/Updates/Update29_data.htm. These data were compiled by Janet Larsen of the Earth Policy Institute from European medical and press sources. Her article about this disaster is at http://www.earth-policy.org/Updates/Update29.htm.

## PART TWO: ECOSYSTEMS IN TROUBLE

### Chapter Six

Kieran Mulvaney's initial research on climate change in the Arctic was conducted during a 3-month voyage in the region onboard the icebreaker *Arctic Sunrise* in 1998; it has subsequently been supplemented through ongoing interviews, research, and writing, much of which was featured in his book *At the Ends of the Earth: A History of the Polar Regions* (Washington, D.C.: Island Press, 2001).

A compilation of Alaska Native observations on climate change was published as *Answers from the Ice Edge: The Consequences of Climate Change on Life in the Bering and Chukchi Seas* by Margie A. Gibson and Sallie B. Schullinger (Anchorage: Arctic Network/Greenpeace, 1998). George Divoky's work on black guillemots in Arctic Alaska was featured in "George Divorky's Planet" by Darcy Frey that appeared as the cover story of the *New York Times Magazine* on January 6, 2002. Two essential overviews of climate change in the western Arctic, including several of the case studies featured in this chapter, were edited by Gunter Weller and Patricia Anderson and published by the Center for Global Change and Arctic System Research of the University of Alaska Fairbanks as *Implications of Global Change in Alaska and the Bering Sea Region* (1997) and *Assessing the Consequences of Climate Change for Alaska and the Bering Sea Region* (1998). The more recent case study of mosquitoes and Brünnich's guillemots was published in 2002 in A. J. Gaston, J. M. Hipfner, and D. Campbell, "Heat and Mosquitoes Cause Breeding Failures and Adult Mortality in an Arctic-Nesting Seabird," *Ibis* 144, pp. 185–191. An examination of the changes in the Bering Sea and western Gulf of Alaska is available in *The Bering Sea Ecosystem* (Washington, D.C.: National Research Council, 1996).

Finally, two recommended and recent overviews include the polar regions chapter of the latest IPCC Assessment Report, *Climate Change 2001: Working Group II: Impacts, Adaptation and Vulnerability*, and a collection of papers in a special issue of the journal *Science* 297, no. 5586, August 30, 2002.

### Chapter Seven

Orna Izakson gathered material for this chapter beginning in 2000, while on assignment for *E/The Environmental Magazine.*

As undergraduates, Rafe Sagarin and Sarah Gilman published their findings on species shifts on the edge of Monterey Bay in "Climate-Related, Long-Term Faunal Changes in a California Rocky Intertidal Community," *Science* 267, 1995. An update appeared in "Climate-Related Changes in an Intertidal Community over Short and Long Time Scales," *Ecological Monographs* 69, 1999.

Sagarin's website, http://www.stanford.edu/~sagarin/HistoricalEcol.html, shows photos of the area, with a nifty feature showing changes in algae cover on the rocks as you pass your cursor over it.

John McGowan's stark report on the zooplankton drop in the California Current off southern California first appeared in "Climatic Warming and the Decline of Zooplankton in the California Current," *Science* 267, 1995. His proposition that warming-related deepening of the thermocline led to the die-off appears in "The Biological Response to the 1977 Regime Shift in the California Current," *Deep-Sea Research* 50, no. 11, 2003. The sooty shearwater's decline is documented in two studies by McGowan and Richard Veit: "Ocean Warming and Long-Term Change of Pelagic Bird Abundance within the California Current System," *Marine Ecology Press Series*, no. 139, 1996, and "Apex Marine Predator Declines 90 Percent in Association with Changing Oceanic Climate," *Global Change Biology* 3, 1997.

More information about the habitats and restoration of San Francisco Bay NWR is available on the refuge's website at http://desfbay.fws.gov. Further data on the endangered California clapper rail and salt marsh harvest mouse may be found at the U.S. Fish and Wildlife Service's website at http://endangered.fws.gov.

Background on the decline of other California marine species is documented in *California's Living Marine Resources: A Status Report, California Department of Fish and Game Publications*, SG01-11, 2001. General information about climate change globally appears in *Climate Change 2001: Third Assessment Report of the Intergovernmental Panel on Climate Change.*

## Chapter Eight

David Helvarg combined reporting trips to the Florida Keys in 2000 and Australia and Fiji in 2001 to examine the impact of climate on the world's reefs. Billy Causey is manager of the Florida Keys National Marine Sanctuary. His sanctuary's website is http://www.fknms.noaa.gov, although it is more fun to visit the site in person. Aquarius, the world's only underwater laboratory, can be observed at http://www.uncw.edu/aquarius. When scientists are living aboard, the website provides live images from the habitat. Reef Relief is one of the more effective environmental groups working to protect the Florida Keys (and the world's) living coral reefs; go to http://www.reefrelief.org.

On the Great Barrier Reef, Jeremy Tager remains an active voice for environmental protection as an advisor to the Australian Democrats and a leader of the Queensland Conservation Council (online at http://www.qccqld.org.au). The Australian Institute of Marine Science in Townsville continues to do leading studies on the impact of climate change on coral. It can be found at http://www.aims.gov.au. Also in Townsville is the Great Barrier Reef Marine Park Authority (online at http://www.gbrmpa.gov.au).

Fiji is one of forty-three member nations of the Alliance of Small Island States (AOSIS) fighting for more rapid global action on reducing greenhouse gas emissions and transitioning to noncarbon renewable energy before their lands and peoples are further damaged, displaced, or destroyed. For more on this alliance, go to http://www.sidsnet.org/aosis. In Fiji, the ministry of local government, housing, squatter settlement and environment has its hands full just trying to keep up. Its efforts may be researched at http://www.fiji.gov. A center for climate and coral research in Fiji is the University of the South Pacific (online at http://www.usp.ac.fj). Also working to link political and environmental rights for Fijians of all ethnic groups is the Fiji Human Rights Commission (online at http://www.humanrights.org.fj).

### Chapter Nine

Sally Deneen gathered material for this chapter starting in 2000, but mainly in 2003, while on assignment in her home region, the Pacific Northwest.

Scientists at the Climate Impacts Group of the University of Washington continue to research how the region's snowpack, hydropower, salmon, forests, and other mainstays are impacted by climate change, and their continually updated work and past publications are posted online at http://tao.atmos.washington.edu/PNWimpacts/Infogate.htm. Abstracts of some papers and scientific presentations referenced in this chapter—including sockeye salmon's woes, the region's shrinking snowpack and floods—can be found online, among others presented at the 2003 Georgia Basin/Puget Sound Research Conference held in Vancouver, British Columbia, at http://www.psat.wa.gov/Publications/2003research/RC2003.htm.

The reader can learn more about the incredible shrinking glaciers of North Cascades National Park and research there at the following websites: http://www.nichols.edu/departments/glacier/ and http://www.nps.gov/noca/massbalance.htm. Historic photos of glaciers are found in "Glaciers of the Conterminous United States" by Robert M. Krimmel (since retired), at http://pubs.usgs.gov/prof/p1386j/us/westus-lores.pdf. The early blooming of lilacs and honeysuckle is reported on in "Changes in the Onset of Spring in the Western United States," *Bulletin of the American Meteorological Society*, March 2001, p. 399.

General information about the impact of climate change on the region is available in "Impacts of Climate Change: Pacific Northwest" by University of Washington scientists Philip Mote and Nate Mantua (online at http://jisao.washington.edu/PNWimpacts/Publications/es.pdf). It is also discussed in "In Hot Water: A Snapshot of the Northwest's Changing Climate" by Patrick Mazza of Climate Solutions (online at http://www.climatesolutions.org/pubs/inHotWater.html).

Deneen, whose reporting on the effects of climate change in Florida won the Charlie Award (first place) for in-depth reporting from the Florida Magazine Association in 2002, wishes to acknowledge Seattle-based environment reporter Robert McClure, her husband, for assistance with the hydropower and irrigation portions of this chapter.

### Chapter Ten

David Helvarg traveled to Antarctica in 1999 and 2000 is part of a National Science Foundation program that allows several professional journalists to visit there each year. For more on NSF's Antarctic work, go to its Office of Polar Programs website at http://nsf.gov/od/opp.

For more of the author's reporting on Antarctica, see chapter 6, "A Rising Tide," in his book *Blue Frontier: Saving America's Living Seas* (New York: Owl Books, 2002).

When not on the ice (during the austral winter), penguin scientist Bill Fraser can be contacted at bfraser@3rivers.net. Giant petrel scientist Donna Patterson may also be reached at patterdo@3rivers.net.

The fossil-fuel-driven climate change they are observing on the Antarctic Peninsula is the most significant but not the only human threat to Antarctica's environment. For more than a quarter century, the Antarctic and Southern Ocean Coalition (ASOC), made up of some 140 nongovernment organizations in over 40 nations, has been working to protect the world's last wild continent from environmental harm. It can be contacted through its website at http://www.asoc.org.

# About *E Magazine*

*E/The Environmental Magazine* debuted in 1990 while the world was celebrating the twentieth anniversary of Earth Day, yet reeling from a series of environmental shocks, including the *Exxon Valdez* oil spill, "greenhouse summers," fires in Yellowstone Park, and medical waste washing up on America's eastern shores. In the time since, *E* has established itself as the leading independent environmental journal.

Edited for the general reader but also of sufficient depth to appeal to the dedicated activist, *E* is a clearinghouse of information, news, commentary, and resources on environmental issues. *E* was founded and is published by Connecticut residents Doug Moss and Deborah Kamlani, and it is edited by Connecticut resident and long-time writer, author, and radio host Jim Motavalli. *E* is a project of the not-for-profit Earth Action Network, Inc., which also owns and manages the environmental website http://www.emagazine.com, where an extensive archive of *E* stories is maintained. *E* also produces the syndicated question-and-answer column "Earth Talk" that appears weekly in a variety of hometown newspapers around the United States.

*E* covers everything environmental—from recycling to rainforests, and from the "personal to the political"—and reports on all key and emerging issues, providing extensive contact information so readers can investigate topics further or plug into activist efforts.

*E* also follows the activities and campaigns of a broad spectrum of environmental organizations, and provides information on a range of lifestyle topics—food, health, travel, "house and home," personal finance, consumer product trends—as they relate to environmental quality.

*E* has drawn considerable recognition since its launch, garnering a dozen awards and citations for its style and content. In 2002, *E* received three *Utne* Independent Press Award nominations and won in the category of "Best Science/Environment Coverage." In 1999, The Population Institute awarded *E* a "Global Media Award" for "Excellence in Population Reporting."

A subscription is $19.95 per year (six bimonthly issues) and can be ordered by writing *E Magazine*, P.O. Box 2047, Marion, OH 43306; by calling 800-967-6572; or by visiting *E* online at http://www.emagazine.com.

# Index


Abbey, Edward, 130
adelie penguins, 158–160, 162–165
  *Chysanoessa Macrura* and, 160
  diet sampling of, 158
  extinction projections for, 163
  study of, 158–160, 162–165
*Aedes aegypti* (mosquito), 76
  Dengue fever and, 76
aerosols, 85, 88
  cooling effects of, 88
  as part of Asian brown cloud, 85, 88
*Against the Tide: The Battle for America's Beaches* (Dean), 48, 56, 178
  shoreline stabilization efforts in, 56
*The Age* (Spillius), 86
AIMS (Australia Institute of Marine Sciences), 130, 132–133
Alaska, 95–102
  annual average temperatures in, 106
  geographic history of, 95–96
  global warming, effects in, 105
  Little Diomede Island, 96–98
  species redistribution in, 99–100
  vegetation changes in, 99
Alaska Climate Research Center, 106
Alliance of Small Island States
  Caribbean islands and, 76
alternative energy sources, 2, 21–22, 30–32
  in China, 21–22
  renewable, 2
  Western Europe, investment in, 30
  wind power as, 31–32
Ambrosini, Monica, 37
American Meteorological Society, 45
The American Littoral Society, 47
Ammerman, Albert, 36–37
  on flooding (Venice, Italy), 36
  on New Venice Consortium, 36–37
Antarctica, 157–171
  adelie penguins in, 158–160, 162–165
  glacier size in, 161
  global warming, effects of, 160–161
  "GPs" in, 165–168
  mean temperature rise in, 161
  meteorological changes in, 161
  research facilities on, 159–160
  size of, 159
anthropogenic (human enhanced) climate changes, 163
  global warming and, 163
Antigua, 61–70, 75
  average income in, 63
  beach sand removal in, 75
  drinking water, availability of, 66
  geography of, 64–65
  geological zones in, 65
  history of, 67
  hurricane damage in, 61–62, 70
  latitude of, 63
  sea-level rise, effect on, 69
  tropical diseases, increase in, 75–76
  vegetation changes in, 65–66
  warming trends in, 66
AOSIS (Alliance of Small Island States), 181
Arctic region, 98, 101, 107
  global warming in, 98, 101
  sea ice, reduction in, 98, 107
*Arctic Sunrise* (boat), 100

Army Corps Beach Erosion Control Project, 53
  New Jersey beaches and, 53
Arthur, Chester A., 51
Asia, 79–92
  brown cloud in, 85–89
  India, 79–85
Asian brown cloud, 85–90, 179
  aerosols as part of, 85, 88
  causes of, 86
  components of, 87
  deaths as result of, 87
  dust movements and, 89–90
  meteorological effects of, 88
  oceanic effects of, 88
  research material for, 179
  size of, 85
  and sunlight, effect on, 88
  UNEP and, 87
"The Asian Brown Cloud: Climate and Environmental Impacts," 87
"The Asian Dust Events of April 1998," 90
*Asperigillus* (fungus), 90
*Atalopedes campestris* (sachem skipper butterfly), 145–146
  climate changes, effect on, 145–146
  in Pacific Northwest, 145–146
Australia, 129–134, 140
  climate-enhanced storms in, 132
  coral "bleaching" in, 129–134
  Great Barrier Reef in, 130–134
Australia Institute of Marine Sciences. *See* AIMS
auto rickshaws, 80
  in India, 80

Baalu, T.R., 92
  on India pollution, 92
Barbuda, 70–75
  coastal management, prospects of, 75
  Codrington, Christopher, and, 71
  coral reefs in, 73
  flora in, 73
  geographic location of, 70
  geological history of, 70–71
  main industries in, 73
  population of, 71
  sand mining, effects of, 72
  species redistribution in, 73
  tourism development in, 74–75
Barnes, Sherry, 173
Baxter, Chuck, 113–114, 117–118, 125
  Gilman, Sarah, and, 113
  Sagarin, Rafe, and, 113
  on *serpulorbus* overgrowth, 113–114
  on tidepool studies, 117
beach replenishment, 47–48, 54
  in *The Corps and the Shore* (Dixon/Pilkey), 48
  in New Jersey, 47–48, 54
*The Beaches Are Moving* (Kaufman/Pilkey), 56, 178
Beijing Energy Efficiency Center, 21
Bennett, Dery, 47, 54–55
  on beach replenishment, 54–55
Berlusconi, Silvio, 37
Bhatia, S.P. Singh, 80
"big gates" project (Venice, Italy), 35–37
  criticism of, 36
  New Venice Consortium and, 35–36
  support for, 37
Blair, Tony, 5
  on carbon emissions (UK), 5
Bloomfield, Janine, 42
Bodman, Noreen, 48
  on beach replenishment, 48
Braasch, Gary, 173
Bradley, Ed, 75
Bras, Rafael, 35
Brown, Hartley, 146
  on glacier shrinkage, 146
Browne, John, 30
Buliruarua, Leigh-Anne, 137
*Bulletin of the American Meteorological Society*, 144
Bush, George H.W., 7
Bush, George W., 7–8
  on global warming, 7
  on Kyoto Protocol, 7–8

CalCOFI (California Cooperative Oceanic and Fisheries Investigations), 119–120, 122
  data collection of, 119
  on ocean temperatures, 120
California coastline, 111–125
  California Current and, 119–122
  ecological zones in, 111–112
  Monterey Bay, 114–118
  San Francisco Bay, 123–125

California Current, 119–122
along California coastline, 119–122
direction of, 119
*Leuroglossus stilbius* in, 120–121
pelagic seabirds in, 121
*Stenobranchus leucopsarus* in, 121
water temperatures in, 120, 122
zooplankton numbers in, 120–121
Canada, 102–104
global warming, effects in, 104
salmon population in, 150–153
species redistribution in, 102
carbon levels, 1–2
effect on global warming, 1–2
Caribbean islands
Alliance of Small Island States and, 76
coral reefs in, 76
dust movements, effect on, 89
"cascade hypothesis," 105
NRC and, 105
Causey, Billy, 127–128, 134, 181
in Florida, 127–128
Cayan, Daniel R., 144
Cecconi, Giovanni, 35
Center for Global Change and Arctic System Research, 106
on climate change, in Alaska, 106
Central Pollution Control Board (India), 80
on nationwide rates, of pollution, 80
Cesar, Herman, 84
on ocean temperatures, 84
Chadwick, Virginia, 133
Challenger, Brian, 66, 75–76
on health trends, in Antigua, 75
on warming trends, in Antigua, 66
Chandler, William, 19
Chaudhry, Mahendra, 135, 140
in Fiji, 135
China, 3, 11–24
advanced technology in, 18–19
air pollution in, 11–13
average incomes in, 16–17
Beijing, 11–13
climate changes, effect on, 15
coal consumption in, 11–12, 14–15, 18, 22
Communist Party in, 23
current health concerns in, 15
death rates, factors for, 15
Deng Xiaoping in, 16
development paradox in, 24
economic goals of, 23–24
energy alternatives in, 21–22
energy consumption in, 18–19, 22
energy efficiency investments in, 17–19, 21
on global warming, official response to, 22
greenhouse gases, emissions in, 18, 20
"honeycomb" stoves in, 12
public transportation in, 12
statistical manipulation in, 20
subsidies in, 18
world environmental future, effect on, 17
China Energy Project, 19–20
*Chysanoessa Macrura* (krill), 160–161
adelie penguins and, 160
global warming, effect on, 161
Ciorra, Anthony, 53–54
Clapton, Eric, 65–66
Antigua and, 65–66
climate changes, 1–2, 5–6, 40, 82–84, 163
accelerating rates of, 1
anthropogenic, 163
China, effect on, 15
corporate response to, 6
in India, effects of, 82–84
in New York City, 40
World Economic Forum on, 5
Climate Impacts Group, 154
on climate changes, 154
Climate Justice Summit (India), 83–84
on global warming, 84
organization of, 83
climate models, 29, 168
for global warming, 168
in Western Europe, 29
Clinton, Bill, 82
$CO_2$ (carbon dioxide), 2, 4, 161
annual mean temperatures, effect on, 2
atmospheric levels of, 2, 161
effect on plant metabolism, 4
coal
in China, consumption of, 11–12, 14–15, 18, 21
in China, production of, 18
as fossil fuel, 1
*Coastal Management*, 56
*coccolithophore* algae, 104
Codrington, Christopher, 71
Barbuda and, 71

Communist Party (China), 23
coral "bleaching," 69, 84–85, 127–131, 133, 138
- in Australia, 129–134
- Coral Reef Task Force on, 131
- Fabricius, Katherina, on, 133
- in Fiji, 138–139
- in Florida, 127–128
- in Great Barrier Reef, 131, 133

Coral Reef Task Force (U.S.), 131
- on coral "bleaching," 131

coral reefs, 69, 73, 76, 127–140
- in Antigua, 69
- in Barbuda, 73
- "bleaching" of, 69, 84–85, 127–131, 133, 138
- in Caribbean, 76
- fishing industry, effect on, 69
- global warming, effects on, 69
- tourist values of, 140

*The Corps and the Shore* (Dixon/Pilkey), 48, 178
- on beach replenishment, 48

CRAB (Citizens Right to Access Beaches), 52
- in New Jersey, 52

Crosier, Lisa, 145–146
Crutzen, Paul, 87
*Culex pipens* (mosquito), 45–46
- West Nile encephalitis and, 45–46

Danish Wind Manufacturers Association, 31
Dare County Project, 55
Day, Tad, 168
- on global warming, 168

Dean, Cornelia, 48
Delta Project, 32–33
- in Netherlands, 32–33
- scope of, 32–33

Deneen, Sally, 173, 182
- research sources for, 182

Deng Xiaoping, 16, 18
- economic reforms of, 16, 18

Dengue fever, 76–77
Denmark, 31–32
- energy sources, percentages for, 31
- on greenhouse gas emissions, 32
- transportation policies in, 31–32
- wind power in, 31

Despommier, Dickson, 46
- on West Nile encephalitis, 46

Devlin, Bob, 50
Divoky, George, 101–102
- species research of, 101–102

Dixon, Katherine, 48
Don Edward's San Francisco Bay National Wildlife Refuge, 123
Douglass, Scott, 54
- on beach replenishment, 54

Duchin, Melanie, 98
dust movements, 89–90
- Asian brown cloud and, 89–90
- Caribbean islands, effect on, 89
- worldwide effects of, 90

*Earth Odyssey: Around the World in Search of Our Environmental Future* (Hertsgaard), 177
*E/Environmental Magazine*, 3, 183–184
- press awards for, 184

Egan, Timothy, 143
El Niño, 84, 120
- ocean temperatures and, 120
- zooplankton, effect on, 120

*Endocladia muricata* (algae), 116–118
- Gilman, Sarah, studies of, 116–118

energy efficiency, 17–21
- China Energy Project and, 20
- in China, investments in, 17–19, 21

energy management corporations, 21
- World Bank and, 21

Environmental Awareness Group (Antigua), 64, 71
Epstein, Paul, 46
- on West Nile encephalitis, 46

E.U. (European Union), 27
Evans, Alex, 27
"extinction pulse," 164
- global warming as, 164

Exxon Mobil, 3–4
- "greenhouse skeptics" and, 3–4

*Exxon Valdez* (ship), 99, 183

Fabricius, Katherina, 133
Fiji, 134–140
- Chaudhry, Mahendra, in, 135
- climate changes in, 136–137
- coral "bleaching" in, 138–139
- global warming, effects of, 136–137
- Gordon, Arthur, in, 134

history of, 134–135
industry in, 135
political instability in, 140
rising sea levels in, 136–137
saltwater intrusion in, 136–137
tourism value in, 140
Fiji Human Rights Commission, 137
*Fiji Post,* 137
"Fishing in Troubled Waters" (Sridhar), 84
Florida Keys (U.S.), 127–129, 140
onshore development in, effects of, 128
pollution in, effect of, 129
reef study in, 129
tourism value of, 140
Follett, Ken, 70
fossil fuels
in Australia, 133
in China, usage in, 11–12
coal, 1
oil, 1
Fountain, Andrew, 142
Fraser, Bill, 158–160, 162–165
on adelie penguins, 158–160, 162–165
Fuller, Eli, 67–70
biographical history, 67
on vegetation, in Antigua, 67–68

Gammon, Richard, 155
Gardin, Paolo, 33–34
in Venice flooding projects, 34
Garfield, James, 51
Gaston, Anthony, 102
GCC (Global Climate Coalition), 6–7, 30
collapse of, 7
Gelbspan, Ross, 174
Gilman, Sarah, 112, 115–118, 123, 125, 181
Baxter, Chuck, and, 113
*Endocladia muricata,* studies of, 116–118
temperature studies of, 112
tidepool studies of, 116–117
Glacier National Park (Montana), 142
glaciers, size decreases in, 142
Glas, Peter C.G., 26, 30
Global Green Deal, 24
Chinese poverty, effect on, 24
global temperatures, 1–2
$CO_2$, effect on, 1–2
mean rise in, 1
global warming
in Alaska, effects of, 105
in Antarctica, effects of, 160–161
anthropogenic climate changes and, 163
in Arctic region, effects of, 98
Bush, George W.,on, 7
in Canada, effects of, 104
carbon levels, effect on, 1–2
in China, official response to, 22
*Chysanoessa Macrura,* effect on, 161
climate models for, 168
coral reefs, effects on, 69
energy transitions as response to, 5
as "extinction pulse," 164
in Fiji, effects of, 136–137
ground-level ozone and, 45
in "Hot Nights in the City: Global Warming, Sea-Level Rise and the New York Metropolitan Area" (Bloomfield), effects of, 42, 58–59
IPCC on, 98
Jamaica Bay (New York), effects in, 43
Kinney, Pat, on effects of, 44–45
media response to, 4–5
monitoring over time, importance of, 118
in Netherlands (Holland), effects in, 28
in New York City, projected effects, 42, 44–45
in *New York Times,* 4
in North Carolina (U.S.), effects in, 55
preventative measures for, 91
species redistribution as result of, 73, 99–100, 104
in Western Europe, response to, 26–27, 30
Western Fuels Coal Association on, 4
global warming gas, 82
in India, 82
Glynn, Peter, 116
GNP (Gross National Product), 18
*vs.* energy consumed, in China, 18
Gordon, Arthur, 134
in Fiji, 134
Gornitz, Vivien, 41–42
on waterfront construction (New York City), 41
*Gould* (ship), 160
GPS (Global Positioning System), 165
"GPs" (giant petrels), 165–168
in Antarctica, 165–168
climate changes, effects on, 167–168

studies of, 165–168
Grant, Ulysses S., 51
Great Barrier Reef, 130–134, 140
coral "bleaching" in, 131, 133
tourism value of, 140
"Greenhouse Effect, Sea-Level Rise and Barrier Islands" (Titus), 56
greenhouse gases, 18, 20, 32, 76
in China, emissions, 18, 20
in Denmark, emissions reductions, 32
"greenhouse skeptics," 3–4
Exxon Mobil and, 3–4
The Greening Earth Society, 4
Greenpeace, 98
ground-level ozone, 45
global warming and, 45
*Guardian,* 86
Gunn, Anne, 103–104
species study of, 103–104

Harrison, Benjamin, 51
Hartig, Ellen Kracauer, 44
on wetlands regulation, 44
Hayes, John, 86
Hayes, Rutherford B., 51
Helvarg, David, 174, 181–182
research sources for, 181–182
Henriksen, Randall, 39–40
Hertsgaard, Mark, 175
Hewatt, Willis, 114–118
tidepool studies of, 114–118
*Hindustan Times,* 80
on pollution, in India, 80
Hoegh-Guldberg, Ove, 93
"honeycomb" stoves, 12, 19
in China, 12, 18
"Hot Nights in the City: Global Warming, Sea-Level Rise and the New York Metropolitan Area" (Bloomfield), 42, 58–59
global warming effects in, 42, 58–59
Hough, Paul, 131
Huebert, Barry Joe, 88
on aerosols, 88
*huotongs* (Chinese dwellings), 19
hurricanes, 62–63, 70
in Antigua, damage from, 61–62, 70
Hurst, Lionel, 62–63, 65, 77
in Antigua, 62
on vegetation changes, in Antigua, 65
at World Water Forum, 77
Hyatt, Kim, 150–153
on salmon population, in Canada, 150–153

*Ibis,* 102
India, 79–85
auto rickshaws in, 80
Central Pollution Control Board in, 80
climate changes in, effects of, 82–84
Climate Justice Summit in, 83–84
Faridabad, pollution in, 80
"fog" in, 81–82
global warming gas in, 82
Mumbai, pollution in, 79–80
New Delhi, pollution in, 80
poverty in, 79–80
Taj Mahal, effect of pollution on, 80–81
Indian Conservation Institute, 80–81
on Taj Mahal damage, 80–81
*Indian Express,* 82
INDOEX (Indian Ocean Experiment), 87, 92
UNEP and, 87, 92
IPCC (Intergovernmental Panel on Climate Change), 98, 107–108, 122
SCE, in Northern Hemisphere, 98, 107
on SST, in Arctic Basin, 107
on temperature increases, predictions for, 108
terrestrial ecosystems, changes in, 107
Irinaga, Matt, 158–160
on adelie penguins, 158–160
Iron Founders Association (India), 81
*Isis* (boat), 67
Izakson, Orna, 175, 180–181
research sources for, 180–181

Jacob, Klaus, 42
flood projections of, 42
Jamaica Bay (New York), 43–44
ecosystem changes in, 44
effect of global warming on, 43
fauna in, 44
Jamaica Bay Wildlife Refuge and, 43
Kennedy Airport, effect on, 43
marshlands, loss of, 43
Joint Caribbean-Pacific Program for Action on Water and Climate, 76
Joseph, Daven, 66, 71, 75

on coastal management, in Antigua, 75
on warming tends, in Antigua, 66
Juday, Glenn, 99–100

Kamlani, Deborah, 183
Kaufman, Wallace, 56
Keahy, John, 34
Kelly, Brendan, 101
Kincaid, Jamaica, 61
Kinney, Pat, 44–45
on global warming, health effects of, 44–45
Kjaer, Christian, 31
Kolar, Marge, 123–124
Kolker, Alex, 44
Krimmel, Bob, 141–142
Kyoto Protocol, 7–8, 20
Bush, George W., on, 7–8

Labarre, Doc, 169–170
Lal, M., 83
Lal, Ritesh, 138
land erosion
of beaches, 53–54
MEC and, 54
Leach, Robin, 70
Leatherman, Stephen, 49
*Leuroglossus stilbius* (fish), 120–121
in California Current, 120–121
Leyden, Brenden, 50
public trust doctrines and, 50
"Life Under the Asian Brown Cloud" (Hayes), 86
*Lifestyles of the Rich and Famous,* 70
Lippincott, George, 50
Little Diomede Island, 96–98
climate changes in, effects of, 97–98
Cold War, effect on, 96
geographic history of, 96
history of, 96
native population of, 96, 98
permafrost in, 97, 107
littoral drift, 49
Lombardi, Paolo, 36

Mahlung, Clifford, 76
on climate changes, 76
Mani, T.S.S., 85
Mans, Deborah, 52–53
Martin, John, 91
on global warming, preventative measures, 91
Matau, Robert, 137
Mazza, Patrick, 154
McGowan, John, 119–123, 125, 181
on ocean temperature changes, 119
MEC (Metro East Coast) Region, 41, 54
on beach erosion rates, 54
on greenhouse emissions, 41
Mehta, Mahesh Chandra, 81
Middlegrunden Wind Farm, 31
Miles, Edward, 154
Miller, Frank, 103
species study of, 103
Monterey Bay, 114–118
Hewatt, Willis, studies in, 114–116, 118
temperature increases in, 116
Moss, Doug, 183
Motavalli, Jim, 175, 178–179, 183
research sources for, 178–179
Mote, Philip, 144, 147–148
on climate changes, 147
on snowpack declines, 148
Mulvaney, Kieran, 175, 180
research resources for, 180
Mussington, John, 71–74
on development, in Barbuda, 74

Nahata, Rita, 83
Nasome, Epeli, 135–137
on temperature changes, in Fiji, 136
National Environmental Protection Agency (China), 15
*National Geographic,* 88
National Resources Defense Council. *See* NRDC
natural gas, 1
*Nature,* 1, 163
Negi, J.G., 82
Netherlands (Holland), 25–33
canals in, purpose of, 27–28
corporate response in, 30
Delta Project in, 32–33
flood strategies in, 29–30, 32
geography of, 25–26
global warming, effect on, 28
land-use strategies in, 30
"New Lands" in, 28
Petten, 25–26

waterway infrastructure, upgrades of, 26, 28
wind power in, 31–32
Netherlands Water Partnership, 29
New Jersey beaches, 47–58
Army Corps Beach Erosion Control Project and, 53
beach replenishment in, 47–48
CRAB and, 52
Long Beach Township, 50–53
as part of Gateway National Recreation Area, 47
property values in, 49–50
public trust doctrines and, 50
rising sea levels, future projections for, 47
Sandy Hook, 47, 51, 57–58
shoreline access to, 50–53
New Jersey Department of Environmental Protection, 51
New Venice Consortium, 35–36
"big gates" project and, 35–36
Cecconi, Giovanni, and, 35
environmentalists and, 36
New York City, Greater, 39–59
abnormal climate changes in, 40
flooding projections for, 41
health concerns in, 44–45
Jamaica Bay in, 43–44
land subsidence in, 42
New Jersey beaches, 47–58
population of, 40
public works infrastructure of, 40
U.S. Army Corps of Engineers report on, 40
West Nile encephalitis in, 45–46
New York State Department of Environmental Conservation, 44
*New York Times*, 4, 20–21, 43
on global warming, 4
on greenhouse gas emissions, in China, 20
on Jamaica Bay (New York), 43
New York/New Jersey Baykeeper, 47
NOAA (National Oceanic and Atmospheric Administration), 128
Norse, Elliott, 143
North Carolina (U.S.), 55
effect of global warming in, 55
North Queensland Conservation Council, 130
NRC (National Research Council), 105
"cascade hypothesis," 105
"regime shifts" and, 105
NRDC (Natural Resources Defense Council), 20

oceans
Asian brown cloud, effects on, 88
Cesar, Herman, on, 84
El Niño and, 120
phytoplankton in, 120
temperature rise in, 84, 119
zooplankton in, 120
*ocinebra* (whelk snail), 114
Ott, Hermann, 27, 28
ozone levels, 45, 89
in California, 89
global warming, effect on, 45

Pacific Northwest, 141–155
*Atalopedes campestris* in, 145–146
climate changes in, 146
development in, effects of, 149–150
floods, predictions for, 149–150
Glacier National Park (Montana), 142
global warming, economic effects of, 148–149, 154–155
population increases in, projections for, 147–148
salmon in, 143
snowpack decline in, 147
South Cascade Glacier in, 142
species redistribution in, 142
temperature, average rise in, 142
tourism in, importance of, 146
vegetation changes in, 144
"white gold" in, 146
Pacific Northwest Climate Impacts Group, 147
Palahad, Janita, 136
Parrish, David, 89
Patterson, Donna, 158–160, 165–168
on adelie penguins, 158–160
giant petrel studies of, 165–168
Pelto, Mauri, 142
permafrost, 97, 107
on Little Diomede Island, 97, 107
Peterson, David, 144
Pew Center on Global Climate Change, 148
*Philadelphia Inquirer*, 49–50

phytoplankton, 120
in oceans, 120
photosynthesis and, 120
zooplankton and, 120
Pilkey, Orrin, 48, 55–56
on beach erosion, 55–56
on shore retreat, 55
Prosper, Junior, 64–67
on development, in Antigua, 67
on drinking water, in Antigua, 66
Prospero, Joseph, 88–89
Psuty, Norbert, 49
public trust doctrine, 50
shoreline access and, 50

Ramanathan, Veerabhadran, 87–88, 91–92
on Asian brown cloud, 87–88, 91–92
Reagan, Ronald, 56
"regime shifts," 105
NRC and, 105
renewable energy sources, 2
as alternative energy, 2
development of, 2
Riedel, Jon, 142
Rinaldo, Andrea, 37
on "big gates" project, 37
"Rising Waters: Impacts of the Greenhouse Effect for the Netherlands," 177
Dutch flood control plans in, 177
*The Rising Tide: Global Warming and World Sea Levels* (Edgerton), 178
Ronde, John de, 28–29
on waterway infrastructure upgrades (Netherlands), 28–29
Rosenzweig, Cynthia, 43–44
Russell, Dick, 175, 178–179
research sources for, 178–179
Ryan, John C., 142

Sagarin, Rafe, 112–113, 115–118, 123, 125, 181
Baxter, Chuck, and, 113
temperature studies of, 112
tidepool studies of, 116–117
San Francisco Bay, 123–125
development in, effects of, 123–125
endangered species in, 123–124
government support for, 124
restoration of, 124
salt marshes in, 123
Sanchez, Rick, 165
Sandy Hook, (New Jersey), 47, 51, 57–58
civic rehabilitation in, 57–58
environmentalists and, 58
history of, 57
Saunders, Ronald, 68
*Saving America's Beaches* (Douglass), 54, 178
SCE (snow cover extent), 107
IPCC on, change in (Northern Hemisphere), 107
*Schnappers* (boat), 159
*Science*, 2, 76, 98, 105, 112, 120
sea ice, 98, 107–108
benefits of, 108
reduction of, Arctic region, 98, 107
*serpulorbus* (tube snail), 113–114, 116
Baxter, Mark, on, 113–114
overgrowth of, 113–114
Shameem, Shaista, 137
Shinn, Eugene, 90
*A Small Place* (Kincaid), 61
Smith, Ken, 55
sockeye salmon, 150–153
in Canada, 150–153
mortality rates for, 153
spawning behavior of, 151–152
water temperatures, effect on, 151–152
Sommen, Jeroen van der, 29
Soolook, Anthony, Jr., 96–97, 109
South Cascade Glacier (U.S.), 142
decrease in, 142
Spillius, Alex, 86
on pollution, In Asia, 86
Spittlehouse, Dave, 153
Springsteen, Bruce, 51
Sridhar, Lalitha, 84
SST (sea surface temperature), 107
increase in, 107
Stallone, Sylvester, 75
*Stenobranchus leucopsarus* (fish), 121
in California Current, 121
species decline, 121
Surfer's Environmental Alliance, 48

Tager, Jeremy, 130–133, 181
Taylor, Robert, 21
Titus, James G., 56
Topper, Klaus, 9
Trager, Ann, 131

"Twice a River: Rhine and Meuse in the Netherlands," 177
Dutch flood control plans in, 177

U.K. (United Kingdom), 3
carbon emission reductions in, 5
mean temperature changes in, 3
UNDP (United Nations Development Programme), 69, 71
Climate Change Project, in Antigua, 69, 71
on sea levels, 69
UNEP (United Nations Environmental Programme), 87–88, 92
on Asian brown cloud, 87
INDOEX and, 87, 92
Unger, Brian, 48, 51–53, 58
on shoreline access, 48
United Nations Development Programme, 15
U.S. (United States)
Alaska, 95–102
goods consumption in, 17
U.S. Army Corps of Engineers, 40, 44
New York City, report on, 40
on waterfront development (New York City), 44
wetlands project evaluation of, 44
U.S. Board of Geographic Names, 162
"U.S. National Assessment of the Potential Consequences of Climate Variability and Change for the Nation," 41
USG (U.S. Geological Survey), 142, 165
on glacier size, 142

*Venice Against the Sea* (Kehaey), 34, 177
Venice flood control plans in, 177
Venice, Italy, 33–38
"big gates" project in, 35–36
carbon-dating in, 36
flood control projects in, 34–35
flooding in, 33–34
geography of, 33–34
government response in, 34
population decrease in, 33
sewer system in, 36–37
Viner, David, 91

*Washington Post,* 20
West Nile encephalitis, 45–46
cases, in U.S., 46
*Culex pipens* and, 45–46
Despommier, Dickson, on, 46
Epstein, Paul, on, 46
in New York City, 45–46
spreading factors for, 45–46, 46
Western Europe, 25–39
alternative energy, investments in, 30
flood plains in, 29
global warming, response to, 26–27, 30
Netherlands, 25–33
Venice, Italy, 33–38
Western Fuels Coal Association, 4
on global warming, 4
Weston, Scott, 149–150
flooding predictions of, 149–150
"white gold," 146
Williams, Richard S., 146
Willner, Andy, 48–49
on beach replenishment, 48
on littoral drift, 49
Wilson, Woodrow, 51
wind power, 31–32
as alternative energy source, 31–32
Denmark and, 31
in Netherlands, 31–32
Woodard, Colin, 175, 177
research sources for, 177
World Bank, 15, 21
death rates, in China, 15
energy management corporations, creation of, 21
World Economic Forum, 5–6
on climate change, 5–6
World Water Forum, 62–63, 76
World Wide Fund for Nature, 36
Wuppertal Institute for Climate, Environment, and Energy, 27

Zhou Dadi, 19, 21, 23
Zimmerman, Rae, 42–43
on governmental cooperation, In New York City, 43
Zipf, Cindy, 58
zooplankton, 120–121
in California Current, 120–121
El Niño, effect on, 120
in oceans, 120
phytoplankton and, 120
species decline and, 120